Statistical Analysis Using R Software

NEW INDIA PUBLISHING AGENCY
New Delhi-110 034

About the Author

Dr. Abhiram Dash is an Agriculture Scientist working in the field of Agricultural Statistics as Assistant Professor in the Department of Agricultural Statistics, College of Agriculture, Bhubaneswar, Odisha University of Agriculture and Technology. He obtained his Ph.D Degree on Statistical Modeling and Forecasting and has to his credit published innovative articles, research papers in reputed national and international journals. He has expertise in forecast modeling of crops, sampling techniques and design of experiments.

Statistical Analysis Using R Software

Abhiram Dash
Assistant Professor (Agricultural Statistics)
Department of Agricultural Statistics
College of Agriculture, Bhubaneswar
Odisha University of Agriculture and Technology
Bhubaneswar, Odisha

nipa®
NEW INDIA PUBLISHING AGENCY
New Delhi-110 034

NEW INDIA PUBLISHING AGENCY
101, Vikas Surya Plaza, CU Block, LSC Market
Pitam Pura, New Delhi – 110 034, India
Email: info@nipabooks.com
Web: www.nipabooks.com

For customer assistance, please contact
Phone: + 91-11-27 34 17 17 Fax: + 91-11- 27 34 16 16
E-Mail: feedbacks@nipabooks.com

ISBN: 978-81-94766-81-0

Co-Published with CRC (CRC Press is an imprint of the Taylor & Francis Group, an informa business). Meant for Sales Only in India, Pakistan, Nepal, Sri Lanka and Bhutan. Other countries' distribution rights are vested with CRC.

Composed and Designed by NIPA.

Preface

The book is designed for UG, PG and Ph.D students of all disciplines to gain fundamental knowledge on use of R software for statistical analysis. The book can also be helpful to the personnel engaged in teaching and research for gaining acquaintance with R software. The book provides fundamental concepts of R software and its application. The codes used in R mainly for statistical analysis purpose are vividly explained in the book. The syntax of various mathematical and statistical functions are mentioned alongwith the output obtained by the use of syntax. The concepts in the book have been analysed and presented in a simple and precise manner to the comprehension of the readers. I hope my endeavour would cater to the needs of the readers and render them immense benefit.

I thank M/s New India Publishing Agency, New Delhi for taking up the publication of the book in neat, flawless and presentable form.

I owe my sincere gratitude and dedicate this little effort of mine to my departed parents who always blessed me for academic excellence.

Author

Contents

1

Introduction to R

R software is a programming language used for statistical analysis, graphical representation and reporting, and is currently developed by the R Development Core Team. R software is available freely under the GNU General Public License, and the binary versions are provided for the Windows, Linux and Mac operating systems. To develop R, effort from all over the world is made in a collaborative manner.

What is GNU?

GNU stands for GNU's Not Unix. It is a Unix like computer operating system, but different from Unix. GNU consists of GNU packages which are program specific and released by GNU project. Free software means that the users of the software have freedom to run and use the software and also can copy, change, improve and distribute the software freely.

Development of R

R software was initially written by Robert Gentleman and Ross Ihaka also known as "R & R" of the statistics department of the University of Auckland, New Zealand. R appeared first in 1993. Since then various contribution to R has been made by a large group of talented individuals by sending code and bug reports. A core group, known as the "R Core Team" modify the R source code archive since 1997. Thus R is a collaborative project with many contributors.

Features of R

- R is a very well-developed, easy to learn and useful programming language which includes several commands and inbuilt functions to carry out the data analysis work. It has also scope for creating user defined functions, loops and conditionals.
- R commands are case sensitive.
- R can effectively handle a large volume of data with high storage facility.

- R uses the data types like numerical, logical, integer, character and complex.
- The data type of the variables used in R need not be declared.
- R provides a large collection of inbuilt functions for data analysis.
- R provides facilities for graphical representation.

Installation of R in Windows operating system

The recent version of R 3.6.2 can be downloaded and installed from the website www.r-project.org and saved in a local directory. The 32-bit version of windows allows the installation of 32-bit version of R, whereas, the 64-bit version of windows allows the installation of both 32-bit and 64-bit version of R. Some of the versions of R released in 2019 are 3.6.2, 3.6.1, 3.6.0 and those released in 2018 are 3.5.3, 3.5.2, 3.5.1, 3.5.0. To download R, the preferred CRAN mirror has to be chosen. CRAN means Comprehensive R Archive Network which is a network of websites that host the R program and that mirror the original R website. It is the main storehouse for R Packages. The preferred CRAN mirror is http://cloud.r-project.org/. After installation of R, R-studio has to be installed for statistical computing. RStudio can be downloaded from the website www.rstudio.org by clicking on Download RStudio button available in the website.

Steps to install R are as follows:

1. After opening an internet browser go to www.r-project.org.
2. The “download R” link found at the middle of the page has to be clicked. On doing so, the CRAN window will be opened.
3. A preferred CRAN location has to be selected and the corresponding link has to be clicked. If the preferred CRAN location is http://cloud.r-project.org/ , then this link has to be clicked. On doing so, a new window will be opened.
4. The “Download R on windows” link in the middle of the page opened window has to be clicked. Now the R will be installed.

After installation of R, R-studio has to be installed for statistical computing. RStudio can be downloaded from the website www.rstudio.org by clicking on Download RStudio button available in the page opened in the website.

CRAN (Comprehensive R Archive Network), is the primary package repository (i.e. central location in which data are stored or managed) in the R community. **CRAN** is a network of FTP (File Transfer Protocol) and web servers around the world that store identical and up-to-date versions of R codes. The packages on the CRAN will be easier to get installed (with install. packages) and will also be tested daily with multiple systems.

RStudio is considered to be an integrated development environment (IDE) for R. R is the programming language whereas, RStudio uses the R language to develop programs. It includes a source, console, environment, history of codes executed, tools for plotting, debugging and workspace management. R and RStudio are inseparable versions of the same program. They cannot be substituted for one another. R may be used without RStudio, but R Studio cannot be used without R. Using Rstudio needs R to be installed. R and RStudio are separate packages. R is the basic package and RStudio is an add-on package that makes R easier to use for beginners. RStudio requires R version 3.0.1 (released in 2013) or higher.

Packages in R

To get more information regarding the contribution to R, the command contributors() can be used. The citation(package = "abc") is used to cite the R package whose name is written in place of abc, in publications. For example to cite the use of package ppcor in R, we have to type citation(package = "ppcor") and press ctrl + enter.

The Output will be:

To cite package 'ppcor' in publications use:

Seongho Kim (2015). ppcor: Partial and Semi-Partial (Part)

Correlation. R package version 1.1.

https://CRAN.R-project.org/package = ppcor

The help () function can be used as help(package = "...") is used for getting help regarding a particular package whose name is written in place of abc. For example, help(package="ppcor") is used for getting help regarding the package ppcor.

R software uses packages which contains functions, to perform its task. To use a function for performing a particular task, the package containing the function must be installed and loaded in the current R session. To quit the current R session, quit() is to be typed.

List of packages already installed in R is found in RStudio window. If a package is not installed, it can be installed by clicking on install button under package menu. Once a package is installed in R, it remains in the hard disk and can be loaded as and when required while running R. But the installed package has to be loaded in every R session either by clicking on the check box at side of the name of the installed package while running R or by typing library("abc") where abc is the name of the package to be loaded.

RStudio window

When the RStudio is opened, it can be seen that it has four windows (also known as panes) shown in Figure1.1. These are the:

(i) source pane/window

(ii) console pane/window

(iii) Environment/history pane/window

(iv) Files/plot/package/help pane

The order of the windows can also be changed under RStudio preferences. The size of the windows can be changed by clicking on minimise or maximise button on the top right of each pane or by clicking and dragging the middle of the borders of the windows.

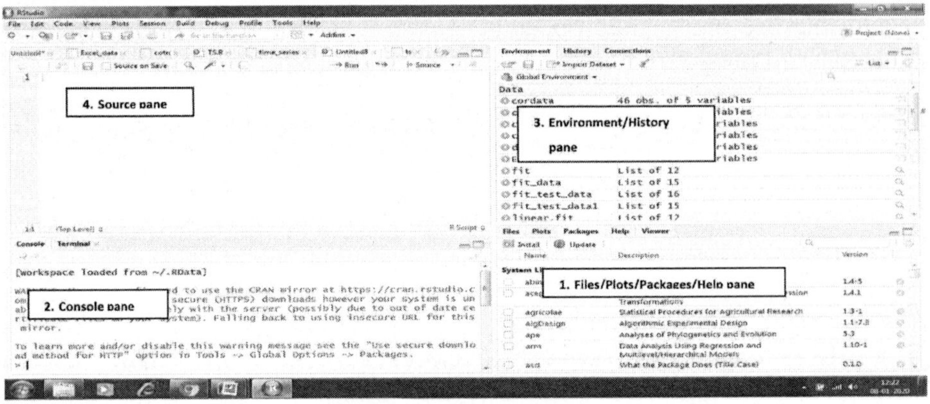

Fig. 1.1: RStudio Window

The description of the four windows/panes is given below:

(i) Source pane: It is the notepad to write the codes. Here the R Scripts can be created and edited. R Scripts are text files with the R extension. R Scripts are the collection of codes in R. When R Studio is opened, a new untitled R Script will automatically be started. To start a new R Script, ctrl+shift+N can be used from keyboard or we can click New File under File menu and click on R Script tool under New File option. Source pane can only be used to type the codes. The codes cannot be executed here. For execution of the codes written in source pane, they have to be transferred to Console pane. This can be done by pressing ctrl + enter or directly clicking on Run button in the top right portion of the Source Pane. The codes written in Source pane is saved in form of a file with a name having extension .R. Usually one statement is written in a line and executed by pressing ctrl + enter. But multiple statements can also be entered in same line where each statement is separated by a semicolon (;). In this case by pressing ctrl+enter, all the statement in the line will be executed at a time. (This will be illustrated in further chapters).

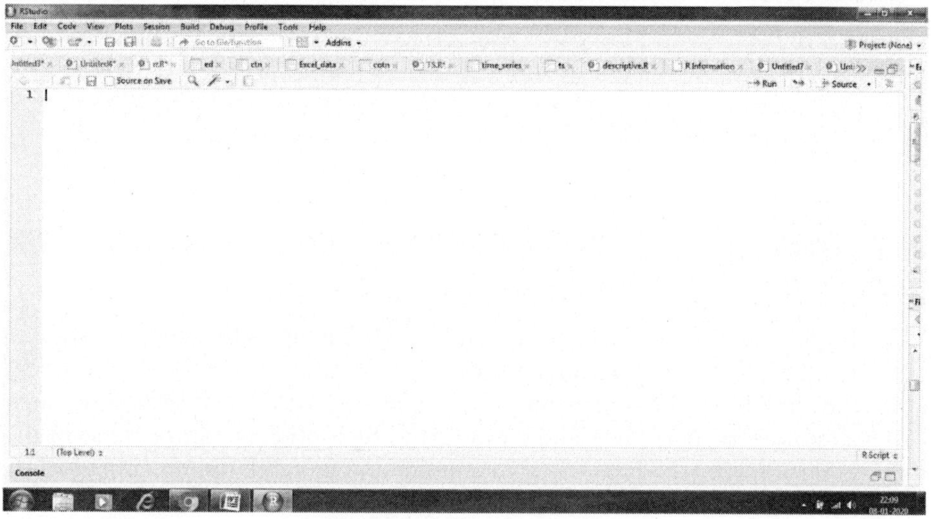

Fig. 1.2: Source pane of R Studio

(ii) Console Pane: The control pane is regarded as the heart of R. The codes of R written in the source pane are actually executed here in the console pane. The R codes can also be directly written in console pane and executed there. But it is usually not done except in some rare cases when we want quick analysis. The problem in writing the codes directly in console pane is that it won't be saved.

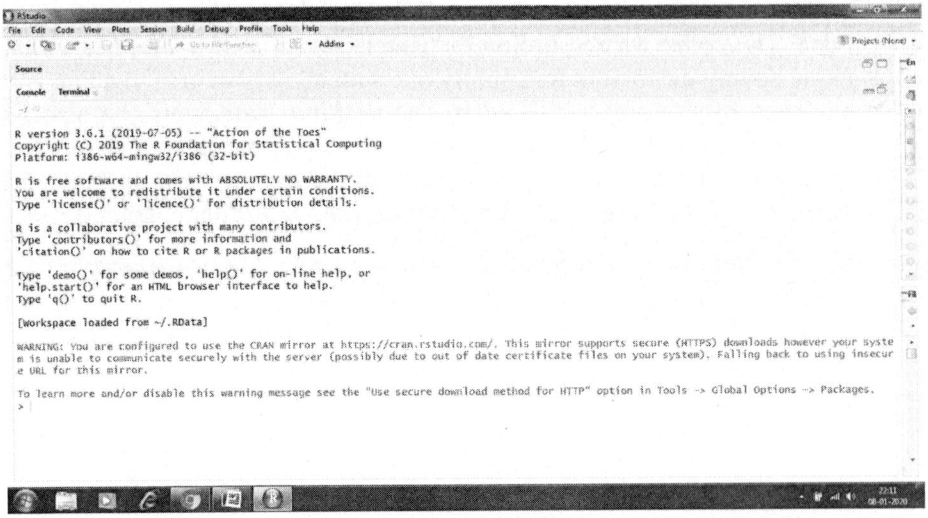

Fig. 1.3: Console panel of RStudio

(iii) Environment/History Pane: The Environment tab of this panel shows the names of all the data objects (like vectors, matrices, lists, data frames) that has been defined in the current R session. The information such as no. of observations and rows in the data objects can be seen in the environment tab. This tab also contain an clickable option 'import data set' which can be used to import data from excel file, text file, csv file, SAS, SPSS and Stata.

The history tab of the panel shows the history of all the codes previously executed in the console pane.

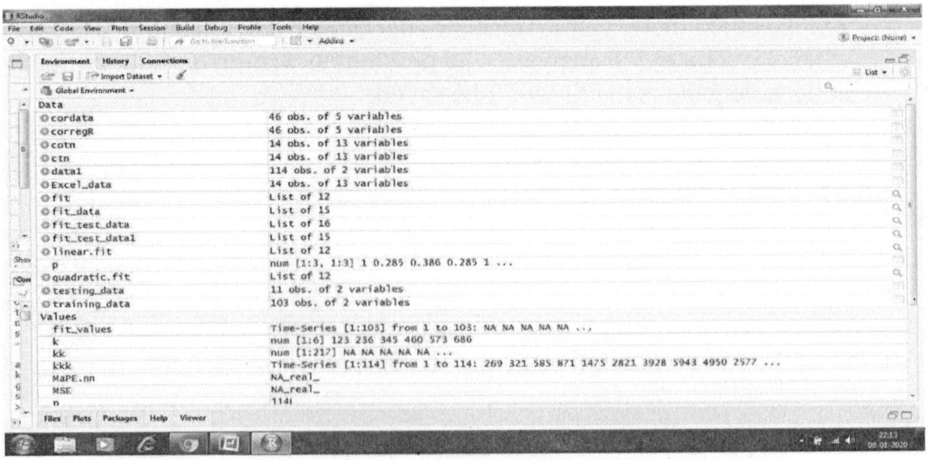

Fig. 1.4: Environment/History pane of RStudio

(iv) Files/Plots/Packages/Help: This pane provides a lot of helpful information. The different tabs under this pane are : (a) Files (b) Plots (c) Packages (d) Help. The description of different tabs are given below:

(a) Files: The files tab of the pane gives access to the files and directories of R in the hard drive. This tab can be used to set the working directory. Working directory is the directory inside which the R file containing the codes in source pane and the output result files can be stored.

(b) Plots: The plots tab shows all the plots obtained as output of R codes for plotting bar diagram, histogram, box plots, scatter plots and other type of plots. Buttons are also available in this pane for opening the plot in a separate window and exploring the plot as a pdf or jpeg which can also be accomplished with code using pdf() or jpeg() functions.

(c) Packages tab: The packages tab shows the list of R packages installed in the hard drive. A particular package can be installed by clicking on install button under package tab. On clicking the install button a small window named as 'install package' is opened where the name of the package (to be installed) has to be written in the given space with blinking cursor and then click on install in the window (Figure 1.6). The package can also be installed by using the code install.packages("abc"), where abc is the name of the package to be installed. The packages that are loaded in the current session are checked and those unchecked are not yet loaded in the current R session. The installed package can be loaded in the current R session by using the code library

(d) Help tab: The help tab of this pane contains help menu which is used for getting help regarding a particular package or function. The code for help in R is help(package="abc") where abc is the name of the package.

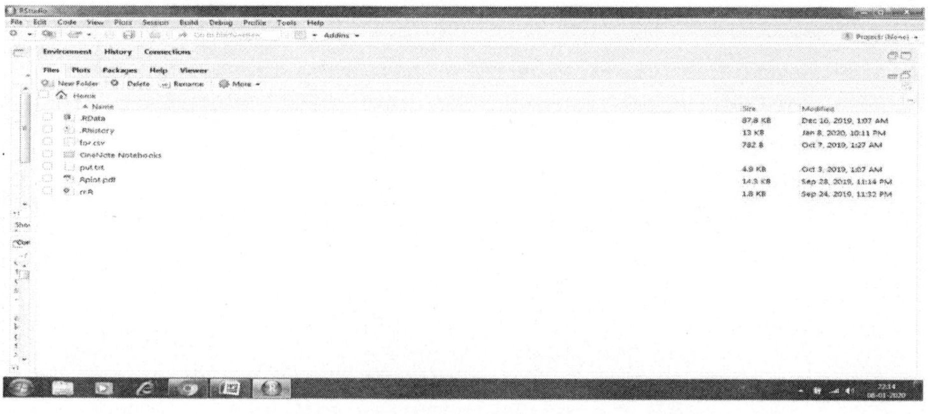

Fig. 1.5: Files/Plots/Packages/Help pane of RStudio

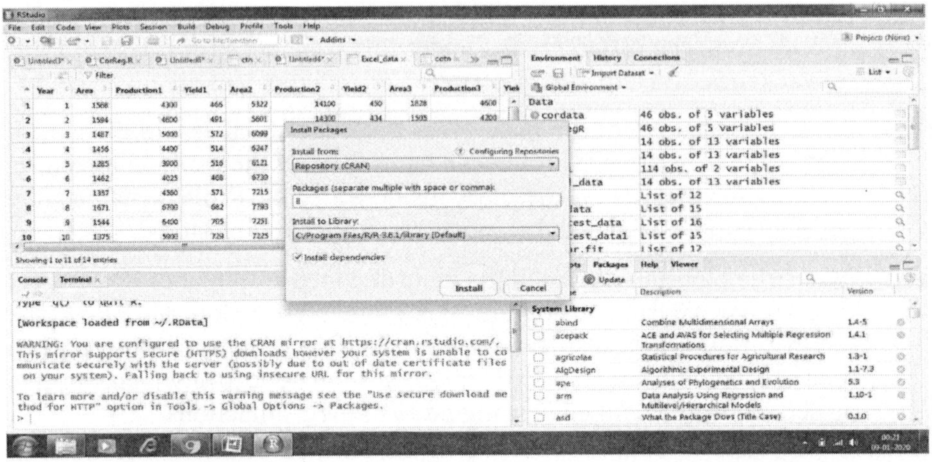

Fig. 1.6: Install package window under packages tab

To execute any statement in RStudio, the ctrl+enter has to be pressed. Usually one statement is written in a line and executed by pressing ctrl + enter. But multiple statements can also be entered in same line where each statement is separated by a semicolon (;). In this case by pressing ctrl+enter, all the statement in the line will be executed at a time

2

Objects, Functions and Operators in R

This chapter deals with the objects used in R which contains the data in various formats. The data types used in R has also been mentioned in this chapter. The chapter also includes the functions used in R as the codes for accomplishing several jobs related to data analysis.

Unlike C language and Java, the variables in R are not declared as some data type. The variables are assigned with R object and the data type of R object becomes the data type of the variable. The commonly used R objects in R are vectors, lists, matrices, arrays, factors and data frames. The vector object is the simplest one and other R-objects are built upon the vectors. The different data types of these vectors are logical, numeric, integer, complex, character and raw. The objects in R are stored by the name of a variable and the data type of the object is determined by its content. To know the data type of a variable the class() function is used. The various data types are described as below:

(i) Logical – Examples of logical data type are TRUE, FALSE.

Illustration:

b = TRUE # *object named as b is assigned a logical value TRUE*

class(b) # *will show the data type of the object named as b*

The output will be

[1] "logical"

b *#will show the value stored in the object named as b*

The output will be

[1] TRUE

The class() function shows the data type of the object.

(The text written in italics after the hash(#) mark denotes the comment of the statement used in R)

(ii) Numeric – For example, 12.34, 5, 4.01, 3.0.

Illustration:

b =12.3*#object named as b is assigned a numerical value 12.3*

class(b)*#will show the data type of the object named as b*

The output will be:

[1] "numeric"

b *#will show the value stored in the object named as b*

The output will be:

[1] 12.3

The numeric data type can assign decimal numbers to a variable.

(iii) Integer – For example, 2L, 4L, 5L

Illustration:

b =3L*# object named as b is assigned an integer value 3*

class(b) *#will show the data type of the object named as b*

The output will be:

[1] "integer"

b *#will show the value stored in the object named as b*

The output will be:

[1] 3

The integer data type can assign only whole numbers to a variable. It cannot assign decimal numbers to a variable. The suffix L after the number is used to save the memory space. If L is not used as suffix then the number would be stored as numeric and this will occupy more memory space. This can be explained much better by use of the following example:

p=1:50

class(p) *#integer*

The output will be:

[1] "integer"

q=p+1

class(q) *#numeric*

The output will be:

[1] "numeric"

object.size(q) *#provides an estimate of memory used to store an R object*

The output will be

432 bytes

(The function object.size() is used to provide an estimate of memory used to store an object in R)

r=p+1L

class(r) # integer

The output will be:

[1] "integer"

object.size(r)

The output will be:

232 bytes

(iv) Complex: Example – 5+ 7i, 3+8i

Illustration:

b=3+4i # *object named as b is assigned a complex number 3+4i*

class(b) *#will show the data type of the object named as b*

The output will be

[1] "complex"

b *#will show the value stored in the object named as b*

The output will be:

[1] 3+4i

(v) Character: Example: "n", "ball", "Tiger"

Illustration:

b= "good" # *object named as b is assigned a character "good"*

class(b) *#will show the data type of the object named as b*

The output will be:

[1] "character"

b # *will show the value stored in the object named as b*

The output will be:

[1] "good"

(vi) Raw: Example- "first" will be stored as 66 69 6e 64

Illustration:

b=charToRaw("first")*#conversion and manipulation of objects of type "raw"*

class(b)*# will show the data type of the object named as b*

The output will be:

[1] "raw"

b*# will show the value stored in the object named as b*

The output will be:

[1] "66 69 72 73 74"

R objects: The R objects contain data of same or different data types (i.e. classes).

The various R-objects are:

(1)Vector: The simplest R-object is the vector which contains one or more elements of a particular data type. The c() function known as combine function is used to combine the elements into a vector. The elements inside the c() function are separated by comma (,).

One illustration of creating vector in R is given below:

city=c('bhubaneswar', 'cuttack', 'puri') *#create a vector with name city*

city # *will show the contents of the vectors named as city.*

The output will be:

[1] "bhubaneswar" "cuttack" "puri"

class(city) # *will display the class of the elements in the vector*

The output will be:

[1] "character"

A vector can have single element or multiple element.

Single element vectors have one element and they are of length one.

Example:

v=(5) *# vector named as v contains an element 5*

v

The output will be:

[1] 5

Multiple element vectors: Contain more than one element. The elements may be of same data type or different. The length of the matrix is the no. of elements in the matrix. Usually the c() function is used to assign multiple elements.

Illustration:

v= c(2,3,4) *# a vector names as v contains elements of integer type*

v

The output will be:

[1] 2,3,4

(Thus here all the elements are displayed as integer type).

v=c(2,3.5,6) *# a vector named as v contains elements of integer and numeric type*

v

The output will be:

[1] 2.0 3.5 6.0

(Thus in combination of numeric and integer data type elements, all the elements is displayed as numeric if at least one element is numeric).

v=c("college",2.6, 4, TRUE) *# a multiple vector named as v in which at least one element is character data type*

The output will be:

[1] "college" "2.6" "4" "TRUE"

(Thus if at least one element in a vector is of character type then all elements of the vector will be character type)

To include a sequence of numbers in a vector the colon (:) operator can be used with numeric data.

Illustration:

v=4:12 *# creating a sequence from 4 to 12*

v

The output will be:

[1] 4 5 6 7 8 9 10 11 12

v=4.2:9.2 *# create a sequence from 4.2 to 9.2*

v

The output will be:

[1] 4.2 5.2 6.2 7.2 8.2 9.2

v=4.2:9.1 *# create a sequence from 4.2 to 8.1*

v

The output will be:

[1] 4.2 5.2 6.2 7.2 8.2

In all the above cases the increment order is 1.

In order to define the increment order or decrement order the sequence operator is used.

Illustration of sequence operator:

v=seq(4,6,by=0.2) *# increment by 0.2*

v

The output will be:

[1] 4.0 4.2 4.4 4.6 4.8 5.0 5.2 5.4 5.6 5.8 6.0

v=seq(6,4,by=-0.2) *# decrement by 0.2*

v

The output will be:

[1] 6.0 5.8 5.6 5.4 5.2 5.0 4.8 4.6 4.4 4.2 4.0

Accessing the elements of vector:

There are various ways of accessing the elements of vector.

(i) Accessing vector elements using positions:

Illustration:

v=c("sun","mon","tue", "wed","thu","fri","sat")

b=v[c(2,5,7)] *#second, fifth and seventh element will be accessed*

b

The output will be:

[1] "mon" "thu" "sat"

(ii) Accessing vector elements using logical indexing:

Illustration:

v=c("sun","mon","tue", "wed","thu","fri","sat")

b=t[c(TRUE,FALSE,FALSE,TRUE,FALSE,TRUE,TRUE)]*#TRUE will fetch the corresponding element*

b

The output will be:

[1] "sun" "wed" "fri" "sat"

(iii) Accessing vector elements by using negative indexing

v=c("sun","mon","tue", "wed","thu","fri","sat")

b=v[c(-2,-3,-6)]# except second, third and sixth element all other elements will be accessed

b

The output will be:

[1] "sun" "wed" "thu" "sat"

Two vectors of same length can be added, subtracted, multiplied and divided to obtain a vector as output.

Illustration:

v1= c(3,4,7,8)

v2=c(2,5.5,6,7)

add.result=v1+v2

add.result

The output will be:

[1] 5.0 9.5 13.0 15.0

Here add.result is just the name of a variable which stores the result of sum of the elements of the two vectors. Any name can be used for the variable which will store the sum. Here more than 2 vectors can also be added but all vectors must contain same no. of elements

Similarly, two vectors can be subtracted as illustrated below.

subtract.result = v1-v2

subtract.result

The output will be:

[1] 1.0 -1.5 1.0 1.0

Multiply.result=v1*v2

Multiply.result

The output will be:

[1] 6 22 42 56

divide.result=v1/v2

divide.result

The output will be:

[1] 1.5000000 0.7272727 1.1666667 1.1428571

If we carry on arithmetic operations on two vectors of unequal lengths then the elements of shorter vector are recycled to complete the operations.

Illustration:

v1=c(2,4,5,6)

v2=c(3,7)

#(Here the elements of vector v2 will be recycled as c(3,7,3,7))

add.result=v1+v2

add.result

The output will be:

[1] 5 11 8 13

But the no. of elements in vector of longer length will be in multiples of the no. of elements in vector of shorter length.

The elements of data type as integer, numeric or character in a vector can be sorted by using sort() function.

Illustration:

v=c(2,3.4,5,6,1,2.3)

k=sort(v)

k

The output will be:

[1] 1.0 2.0 2.3 3.4 5.0 6.0

v=c(2,3.4,5,6,1,2.3,"jj") #combination of elements of numeric and character data type.

k=sort(v)

k

The output will be:

[1] "1" "2" "2.3" "3.4" "5" "6" "jj"

Thus in the above illustration we find that if at least one of the element in the vector is of character data type then while sorting in ascending order all elements (which are even not of character data type) will be of character data type and the elements when sorted, the character data type will placed after the numeric data type.

v=c("apple","animal","delhi","cuttack")# elements of character type

k= sort(v)

k

The output will be:

[1] "animal" "apple" "cuttack" "delhi"

In case of character data type, the sorting will be done in alphabetic order

One more illustration is also given:

vv=c(2,5,4,3,9,"ak","a")

sort(vv)

The output will be:

[1] "2" "3" "4" "5" "9" "a" "ak"

To sort in decreasing order, we have to give the argument decreasing = TRUE inside the sort function.

Illustration:

v=c(2,3.4,5,6,1,2.3)

k=sort(v,decreasing=TRUE)# TRUE is a logical value and it should be in upper case

k

The output will be:

[1] 6.0 5.0 3.4 2.3 2.0 1.0

v=c(2,3.4,5,6,1,2.3,"jj") #combination of elements of numeric and character data type.

k=sort(v, decreasing=TRUE)

k

The output will be:

[1] "jj" "6" "5" "3.4" "2.3" "2" "1"

v=c("apple","animal","delhi","cuttack")# elements of character type

k= sort(v, decreasing=TRUE)

k

The output will be:

[1] "delhi" "cuttack" "apple" "animal"

2. Lists – It is an R-object which can contain elements of different data types, vectors, functions and even another list inside it.

Illustration:

b=c(2,3,4) # create vector named as b

sum(b) # find sum of the elements in the vector b

list1=list(1,2.3,2L,'sun', sum(b), c(2,6,9))# list named as list1

list1 # will display the elements in the list named as list1

The output will be:

[[1]]

[1] 1

[[2]]

[1] 2.3

[[3]]

[1] 2

[[4]]

[1] "sun"

[[5]]

[1] 9

[[6]]

[1] 2 6 9

class(list1)

The output will be :

[1] "list"

A list can contain a vector and another list.

Illustration:

list1=list(c("apple","orange","guava"),list("Hari",54.9,45L))# here list1 is the name of the list

list1

The output will be:

[[1]]

[1] "apple" "orange" "guava"

[[2]]

[[2]][[1]]

[1] "Hari"

[[2]][[2]]

[1] 54.9

[[2]][[3]]

[1] 45

Accessing of elements in a List:

Illustration:

list1=list(c("apple","orange","guava"),list("Har",54.9,45L))# *creation of list containing a vector and a list*

names(list1)=c("fruits", "An inner list")# names() function is used to allot names to corresponding elements in a list.

list1[1]# will display the content of first element of the list named as list1 which is a vector

The output will be:

$fruits

[1] "apple" "orange" "guava"

list1[2] #will display the content of second element of the list named as list1 which is also a list.

The output will be:

$`An inner list`

$`An inner list`[[1]]

[1] "Har"

$`An inner list`[[2]]

[1] 54.9

$`An inner list`[[3]]

[1] 45

list1$fruits # will display the element (named as fruits) in the list named as list1

The output will be:

[1] "apple" "orange" "guava"

Two or more lists can be merged by use of c()

Illustration:

list1=list("sun","mon",2,3.4)

list2=list("apple","mango","guava")

merge.list=c(list1,list2) #Here merge.list is the name of the variable which stores the elements of the two lists.

merge.list

The output will be:

[[1]]

[1] "sun"

[[2]]

[1] "mon"

[[3]]

[1] 2

[[4]]

[1] 3.4

[[5]]

[1] "apple"

[[6]]

[1] "mango"

[[7]]

[1] "guava"

A List can be converted to vector by using unlist() function

Illustration:

list3=list(1.5,6L,3)

list3

The output will be:

[[1]]

[1] 1.5

[[2]]

[1] 6

[[3]]

[1] 3

un3=unlist(list3)# converting the list3 to vector named as un3

un3

The output will be:

[1] 1.5 6.0 3.0

list4=list(1.2,3,4L)

list4

The output will be:

[[1]]

[1] 1.2

[[2]]

[1] 3

[[3]]

[1] 4

un4=unlist(list4))# converting the list4 to vector named as un4

un4

The output will be:

[1] 1.2 3.0 4.0

b=un3+un4# adding the two vectors

b

The output will be:

[1] 2.7 9.0 7.0

3. Matrices: Matrix is a two-dimensional rectangular data set consisting of rows and columns. Matrix is created by using a vector input to a matrix function. An illustration of creating a matrix type of R-object is given below:

A=matrix(c(2,4,5,6,8,9), nrow=3,ncol=2)

A # will display the elements in the matrix named as A

The output will be:

	[,1]	*[,2]*
[1,]	*2*	*6*
[2,]	*4*	*8*
[3,]	*5*	*9*

The syntax for creating a matrix is as follows:

matrix(data,nrow,ncol,byrow)

The description of the parameters is as follows:

data- the input vector which becomes the data elements of the matrix.

nrow – no. of rows to be created.

ncol – no. of columns to be created.

byrow – a logical value which if TRUE then the elements will be arranged as rows. If FALSE then the elements will be arranged as columns.

Illustration:

A=matrix(c(2,3,4,5,6,7),nrow=3,ncol=2,byrow=TRUE)

A # will display the elements in the matrix named as A

The output will be:

	[,1]	*[,2]*
[1,]	*2*	*3*
[2,]	*4*	*5*
[3,]	*6*	*7*

byrow=TRUE will arrange the data rowwise

Illustration:

A=matrix(c 2,4,5,6,8,9,nrow=3,ncol=2,byrow=FALSE)

A # will display the elements in the matrix named as A

The output will be:

	[,1]	*[,2]*
[1,]	*2*	*6*
[2,]	*4*	*8*
[3,]	*5*	*9*

byrow=FALSE will arrange the data columnwise

Thus by default byrow is set to be FALSE

---To add row and column names in the matrix- The argument dimnames=list() is used.

Illustration:

rownames=c("row1","row2","row3")

colnames=c("col1","col2")

M= matrix (c(4,9,3,8,7,2),nrow =3, ncol=2,byrow = TRUE, dimnames = list (rownames,colnames))

M

The output will be:

	col1	*col2*
row1	*4*	*9*
row2	*3*	*8*
row3	*7*	*2*

All the elements of a particular row or column or the element a particular element of any row and column of the matrix A created above can be displayed in the following way:

b=A[1,2]# will display the element of matrix A belonging to row 1 and Col 2

b # will display the output

The output will be:

[1] 9

b=M[,2] # *will display all elements of second column.*

b # will display the output

The output will be:

row1 row2 row3

9 8 2

b=A[2,] # *will display all elements of second row*

b # *will display the output*

The output will be:

col1 col2

3 8

The matrix operations are discussed in details in chapter 6.

(4) Arrays: Matrices are confined to only two dimensions but arrays can have any number of dimensions. The array function takes a dim attribute which creates the required number of dimension. In the below example we create an array with two elements which are 3x3 matrices each.

Illustration:

arr = array(c('green','yellow'),dim = c(3,3,2)) # *Create an array named as arr*

arr

The output will be:

, , 1

	[,1]	[,2]	[,3]
[1,]	"green"	"yellow"	"green"
[2,]	"yellow"	"green"	"yellow"
[3,]	"green"	"yellow"	"green"

, , 2

	[,1]	[,2]	[,3]
[1,]	"yellow"	"green"	"yellow"
[2,]	"green"	"yellow"	"green"
[3,]	"yellow"	"green"	"yellow"

Here 2 no.s of arrays are formed. Each array is a matrix of order 3 x 3

Another illustration:

v1=c(2,7,9,8)# *A vector of length 4 is created*

v2=c(1,8,0,6,5,4)# *A vector of length 6 is created*

col.names=c("c1","c2","c3","c4","c5")

row.names=c("row1","row2")

matrix.names=c("matrix1","matrix2")

array1=array

(

c(v1,v2),dim=c(2,5,2),dimnames=list(row.names,col.names,matrix.names)

)

array1

The output will be:

, , matrix1

	c1	c2	c3	c4	c5
row 1	2	9	1	0	5

row2 7 8 8 6 4

, , matrix2

	c1	*c2*	*c3*	*c4*	*c5*
row1	*2*	*9*	*1*	*0*	*5*
row 2	*7*	*8*	*8*	*6*	*4*

Here col.names is the name of a vector data object which contains the elements, c1,c2,...c5. We can use any name (say, c.name or col_name or columns) in place of col.names. Similar is the case with row.names which is used as the name of a vector data object which contains the elements, row1,row2. We can use any name (say, r.name or row_name or rows) in place of row.names. Similarly, matrix.names which is used as the name of a vector data object which contains the elements, matrix1,matrix2. We can use any name (say, m.name or matrix_name or matrix) in its place. Also, array1 is the name of the variable that contains the output of the function array. Here also we can use any name (say, array or array_name) in place of array1.

The function array() is used to create an array. Inside the function array(), the two vectors are combined by using c() function, the attribute dim is used in form of dim=c(x,y,z) where c(x,y,z) function combines x (i.e. no. of rows in a matrix), y (i.e. no. of columns in a matrix), z (i.e. no. of matrices). The attribute dimnames contain the list data object which will specify the name of the vectors used for rownames, column names and matrix names.

Here one important thing to note is that the order in which x, y and z are mentioned must be in accordance with the order in which the row, column and matrix names are mentioned.

(5) **Factors:** Factors are the data objects which are used to categorise the data and store its levels. They can store both strings and integers. They are useful for statistical modelling. To create a factor, factor() function is used by taking vector as input.

v = c ("rice"," wheat", "maize", "rice", "wheat", "ragi", "maize", "ragi", "rice", "maize")

#create a vector named as v

v

The output will be:

[1] "rice" "wheat" "maize" "rice" "wheat" "ragi" "maize" "ragi" "rice"

[10] "maize"

is.factor(v) # *is.factor() function test whether the data object is a factor or not.*

The output will be:

[1] FALSE

factor_data=factor(v) #*factor() function converts vector to factor data object*

factor_data

The output will be:

[1] rice wheat maize rice wheat ragi maize ragi rice maize

Levels: maize ragi rice wheat

is.factor(factor_data)

The output will be:

[1] TRUE

nlevels(factor_data) # nlevels() functions gives the count of levels

The output will be:

[1] 4

The no. of levels in the factor (named as factor_data) is 4.

The factor levels can be generated by using the gl() function.

The syntax for gl() function is gl(n,k,labels)

The description of the parameters n,k, labels are the following:

n is the no. of levels; k is the no. of replications for each level i.e. no. of times the each level is repeated;

labels is an argument which contains the levels of the factor.

Illustration:

Suppose we have 3 treatments named as A, B C each with 4 replications each.

v=gl(3,4,labels=c("A","B","C"))

v

The output will be:

[1] A A A A B B B B C C C C

Levels: A B C

(6) Data Frames: Data frames are tabular data objects. Unlike a matrix in data frame each column can contain different modes of data. Each column is a vector. The data type of the elements of a vector representing a column may be different from the data type of elements of another vector representing another column of the data frame. Some column contains elements of numeric data type while the other column contains character and other one can be logical. All the vectors representing a column should be of equal length i.e. they contain same no. of elements.. Each column contains values of one variable and each row contains one set of values for each column.

Data frames are created by using the data.frame() function.

Illustration: Let us prepare a data frame named as student_data which contains the columns with headings as names, roll no., marks and DOB

```
student_data=data.frame(
name=c("Hari","Sita","Gopal","Shyam","Radha","Govind"),
roll_no.=c(1:6),
marks=c(94,67,88,98,90,92),
DOB= c ("03.02.2006","06.03.2006","09.01.2006","12.05.2007","14.11.200
5","15.09.2006"))
student_data
```

The output will be:

	name	*roll_no.*	*marks*	*DOB*
1	*Hari*	*1*	*94*	*03.02.2006*
2	*Sita*	*2*	*67*	*06.03.2006*
3	*Gopal*	*3*	*88*	*09.01.2006*
4	*Shyam*	*4*	*98*	*12.05.2007*
5	*Radha*	*5*	*90*	*14.11.2005*
6	*Govind*	*6*	*92*	*15.09.2006*

The structure of data frame can be seen by using the str() function. Here we have to write the name of the data frame within the parentheses.

Illustration:

```
str(student_data)
```

The output will be:

'data.frame': 6 obs. of 4 variables:

$ name : Factor w/ 6 levels "Gopal","Govind",..: 3 6 1 5 4 2

$ roll_no.: int 1 2 3 4 5 6

$ marks : num 94 67 88 98 90 92

$ DOB : Factor w/ 6 levels "03.02.2006","06.03.2006",..: 1 2 3 4 5 6

R Operators: An operator is a symbol used for performing specific mathematical or logical manipulations. There are many built-in operators in R language. The various categories of operators used in R are the following:

- Arithmetic operators
- Relational Operators
- Logical operators
- Assignment operators

Apart from these there are also some miscellaneous operators.

A brief description of the operators used in R language is given below.

1. Arithmetic Operators: The arithmetic operators include the operator for addition, subtraction, multiplication, division, remainder, quotient, exponent.

Addition operator: The symbol for addition operator is '+'. It is used to add two or more numbers or variables (containing a number) or vectors (containing elements of integer or numeric data type).

Illustration1: Use of addition operator '+' to add the numbers without using any variable

5+9

By pressing ctrl + enter after writing 5+9, the output will be:

[1] 14

(Here the output is not stored in any variable) (Note that the output always appears in the console window)

c=5+9

c

The output will be:

[1] 14

(Here the output is stored in the variable named as c)

Illustration 2: Use of addition operator '+' to add the numbers stored in variable

a=5;b=-6

c=a+b

c

The output will be:

[1] -1

Illustration 3: Use of addition operator '+' to add the numbers stored as elements of vectors

v1=c(3,4,-1);v2=c(-2.3,1,6); v3=c(-2.3,1,2)

d=v1+v2+v3

d

The output will be:

[1] -1.6 6.0 7.0

(Here all the vectors have same no. of elements)

Illustration 4: Use of addition operator '+' to add the numbers stored as elements of vectors having different no. of elements but the no. elements in one vector is a multiple of no. of elements in another vector.

v1=c(3,4,5);v2=c(-2,3.4,5,6,1,2)

v1+v2

The output will be:

[1] 1.0 7.4 10.0 9.0 5.0 7.0

(Here the 3 no.s of elements in the vector v1 are added for two times to the 6 no.s of elements in vector v2)

Illustration 5: Use of subtraction operator '-'

3-5

The output will be:

[1] -2

c(2,-3,4)-c(0,2,3)

The output will be:

[1] 2 -5 1

Illustration 6: Use of multiplication operator 'x'

7*2*3

The output will be:

[1] 42

c(3,-1,2)*c(1,3,-2)*c(0,1,-4)

The output will be:

[1] 0 -3 16

Illustration 7: Use of division operator '/'

7/2

The output will be:

[1] 3.5

Illustration 8: Use of division operator '/'

c(-9,3,6)/c(0,-2,3)

The output will be:

[1] -Inf -1.5 2.0

Illustration 9: Use of remainder operator '%%'

45%%5

The output will be:

[1] 0

47%%6

The output will be:

[1] 5

The operator %/% gives the quotient

45%/%5

The output will be:

[1] 9

47%/%5

The output will be:

[1] 9

46%/%7

The output will be:

[1] 6

48%/%6

The output will be:

[1] 8

(2) Relational Operators: These operators are used to compare two or more numeric values. The various relational operators are Greater than(>), Smaller than(<),Greater than or equal to (≥), Smaller than or equal to (≤), equal to (==) and not equal to (!=).

Illustrations:

5<4

The output will be:

[1] FALSE

5==10/2

The output will be:

[1] TRUE

6!=3*2

The output will be:

[1] FALSE

v1=c(2,4,-5); v2=c(-1,2,-3)

v1>v2

The output will be:

[1] TRUE TRUE FALSE

c(2,3)!=c(2,-3)

The output will be:

[1] FALSE TRUE

(Here the corresponding elements of the two vectors are compared)

(3) Logical operators: The types of logical operators in R are the following-

(i) AND (&) operator (ii) OR (|) operator (iii) NOT(!) operator

(i) AND operator compares each element of the first vector with the corresponding element of the second vector. The result of comparison is a Boolean value.

Illustration of AND (&) operator:

c(2,3+2i,TRUE)&c(3,4+1i,TRUE)

The output will be:

[1] TRUE TRUE TRUE

c(2,3+2i,TRUE)&c(3,4+1i,FALSE)

The output will be:

[1] TRUE TRUE FALSE

c(0,3+2i,TRUE)&c(0,4+1i,FALSE)

The output will be:

[1] FALSE TRUE FALSE

(If one of the elements in one or both the vectors are zero, then the output will be FALSE)

(ii) OR (□) operator: The OR (□) operator will return TRUE if an element in any one vector is 0. If an element in both the vectors are 0 then it will return FALSE.

Illustration:

c(0,3+2i,TRUE)|c(2,4+1i,FALSE)

The output will be:

[1] TRUE TRUE TRUE

c(0,3+2i,TRUE)|c(0,4+1i,FALSE)

The output will be:

[1] FALSE TRUE TRUE

(iii) NOT(!) operator: This operator takes each value of a vector and returns the opposite result.

!c(5,3+2i,4,7)

[1] FALSE FALSE FALSE FALSE

z=c(5,3+2i,4,0)

!z

[1] FALSE FALSE FALSE TRUE

(iv)Assignment Operators: These operators assign some values to a variable or vector.

There are two types of assignment operators.

(a) Left assignment operators: These are <- or = or <<-

Illustration:

z1<-4

z1

The output will be:

[1] 4

z2<-c(3,4,5)

z2

The output will be:

[1] 3 4 5

z3<<-c(-2, 3)

z3

The output will be:

[1] -2 3

(b) Right assignment operators: These are -> or ->>

Illustration:

34->z4

z1

The output will be:

[1] 34

c(32,34)->>z5

z5

The output will be:

[1] 32 34

Apart from these four types of operators discussed above there are also some other types of operators which are included under miscellaneous operators. These operators are used for some specific purpose such as to assign a series of consecutive numbers to a vector, to check whether a number belongs to a vector or not, to multiply a matrix with its transpose.

The miscellaneous operators are colon (:) operator which is used to assign a series of consecutive numbers to a vector, %in% operator used to check whether a number belongs to a vector or not, and %*% operator to multiply a matrix with its transpose. The %in% operator returns a Boolean value. If the number belongs to the vector then it returns TRUE and if the number does not belong to the vector then it returns FALSE.

Illustration for colon(:)operator:

v1=c(2:8)

v1

The output will be:

[1] 2 3 4 5 6 7 8

v2=c(2,5,4)

v2%in%3

The output will be:

[1] FALSE FALSE FALSE

v2%in%5

The output will be:

[1] FALSE TRUE FALSE

v2%in%1:5

The output will be:

[1] TRUE TRUE TRUE

v2%in%4:5

[1] FALSE TRUE TRUE

3

R Packages

R packages include functions, codes and sample data. The functions used in R are contained in packages. A package in R contains several functions which are related to some particular type of job. A package conatins at least one function and there is no maximum limit to the no. of functions in a package. The set of packages in R which are set by default in the software is called base package. It contains a lot of commonly used functions. The packages in R are stored under a directory called "library" in the R environment. The packages get installed by default while installing R in the computer. Also more packages can be added later on after installation, when required for some specific purposes. The base packages are available by default when we start using R. But the other packages which are already installed while installing R have to be loaded while using R. These packages can be viewed in the package window by clicking on package. To install the package we have to click on install menu in the package window and then write the name of the package.

The packages can also be installed by using the function install.packages("abc") where abc is the name of the package to be installed.

Once a package is installed, it remains in the hard disk of the PC/laptop. But the package will have to be called for use in the current R session. The package once installed can be loaded for use in the current R session by checking the box near the name of the package displayed after clicking on the install menu in the Files/Plots/Packages/Help window. The package can also be loaded by using library() function. The command will be library("abc") where abc is the name of the package. The loaded package remains in the memory for the current R session. Once we quit the R session, the installed package has to loaded again if needed.

The .libPaths() function will give the library locations containing R Packages.

The output of the .libPaths() function will be:

[1] "C:/Program Files/R/R-3.6.1/library"

(This ouput may depend on the location of the folder in concerned PC/laptop)

The library() function gives the list of all packages installed in your PC/laptop.

The output of library() will be the name and description of all packages installed in your PC/laptop. The name of the package is given alphabetically.

A sample output of the list of installed packages with name and description is given below:

Packages in library 'C:/Program Files/R/R-3.6.1/library':

abind Combine Multidimensional Arrays

acepack ACE and AVAS for Selecting Multiple Regression

Transformations

agricolae Statistical Procedures for Agricultural

Research

AlgDesign Algorithmic Experimental Design

ape Analyses of Phylogenetics and Evolution

arm Data Analysis Using Regression and

Multilevel/Hierarchical Models

asd What the Package Does (Title Case)

askpass Safe Password Entry for R, Git, and SSH

assertthat Easy Pre and Post Assertions

aTSA Alternative Time Series Analysis

backports Reimplementations of Functions Introduced Since

R-3.0.0

base The R Base Package

base64enc Tools for base64 encoding

BDgraph Bayesian Structure Learning in Graphical Models

using Birth-Death MCMC

bestglm Best Subset GLM and Regression Utilities

BH Boost C++ Header Files

bibtex Bibtex Parser

boot Bootstrap Functions (Originally by Angelo Canty for S)

brew Templating Framework for Report Generation

callr Call R from R

car Companion to Applied Regression

carData Companion to Applied Regression Data Sets

cellranger Translate Spreadsheet Cell Ranges to Rows and Columns

checkmate Fast and Versatile Argument Checks

class Functions for Classification

classInt Choose Univariate Class Intervals

Apart from the existing packages in the R library, several other packages can also be developed as per the requirement. The R package must contain at least one function. The procedure for development of new package is beyond the scope of the book.

4

Merging of Data in R

In R there is facility of reshaping the data. Reshaping of data includes joining of two or more vectors to create a data frame. Also two or more data frames can be merged. Appropriate functions are available for these jobs. While merging the data frames, we can join columns or rows.

Illustration of joining two vectors by using cbind() function.

```
v1=c(3,4,5,6)
v2=c(12,16,20,24)
j=cbind(v1,v2)
j
```

The output will be:

```
     v1 v2
[1,]  3 12
[2,]  4 16
[3,]  5 20
[4,]  6 24
```

Another illustration of joining three vectors by using cbind() function:

```
Roll_No.<-c(1,2,3)
Marks_english=c(38,48,44)
Marks_maths=c(42,49,46)
student_data=cbind(Roll_No.,Marks_english,Marks_maths)
student_data
```

The output will be:

Roll_No. Marks_english Marks_maths

[1,] 1 38 42

[2,] 2 48 49

[3,] 3 44 46

We can join rows of two or more data frames having identical columns. This is done by using rbind() function. Also columns of two or more data frames can be joined provided that the no. of rows for the columns of different data frames to be merged remains same. Here cbind() function is to be used to accomplish the job.

Illustration of joining rows of two data frames having identical columns by using rbind() function.

#creation of first data frame

Employee_data=data.frame(

emp_code=c(12,23,43),

Designation=c("officer","clerk","peon"),

Salaray=c(80000,60000,40000)

)

The output will be:

emp_code	*Designation*	*Salary*
1	*12 officer*	*80000*
2	*23 clerk*	*60000*
3	*43 peon*	*40000*

#creation of second data frame

Employee_data1=data.frame

(

emp_code=c(65,25),
Designation=c("Officer(A)","Assistant"),
Salary=c(90000,50000)

)

Employee_data1

The output will be:

emp_code	*Designation*	*Salary*
1	*65 officer(A)*	*90000*
2	*25 Assistant*	*50000*

print("Please merge the rows of two data frames")

The output will be:

[1] "Please merge the rows of two data frames"

#merging of rows of two data frames

Merged_row=rbind(Employee_data,Employee_data1)

Merged_row

The output will be:

emp_code	*Designation*	*Salary*
1 12	*officer*	*80000*
2 23	*clerk*	*60000*
3 43	*peon*	*40000*
4 65	*officer(A)*	*90000*
5 25	*Assistant*	*50000*

Illustration of joining columns of two data frames having equal no. of rows by using cbind() function.

#creation of first data frame

Employee_data=data.frame(

emp_code=c(012,023,043),

Designation=c("officer","clerk","peon"),

Salary=c(80000,60000,40000))

Employee_data

The output will be:

emp_code Designation Salary

1 12 officer 80000

2 23 clerk 60000

3 43 peon 40000

#creation of second data frame

Employee_data2=data.frame(

status=c("On tour", "Absent", "Present")

)

#merging of rows of two data frames

Merged_column=cbind(Employee_data,Employee_data2)

Merged_column

The output will be:

emp_code	*Designation*	*Salary status*
1	*12 officer*	*80000 On Tour*
2	*23 clerk*	*60000 Absent*
3	*43 peon*	*40000 Present*

5

Import and Export of Files in R

The most important use of R programming lies in importing of data files which serves as input for the analysis work. After the analysis is over by using R codes, operators and functions, the output result must also be exported in form of files. The files are imported in form of csv files (comma separated variable), text files or excel files.

Before going to the actual content of the chapter we have to have an idea about csv files. CSV stands for Comma Separated value. It is used to store tabular data, such as a spreadsheet or database. CSV is delimited text file in which the values are separated by commas. CSV files occupy much less space than excel files and also gets imported in a program at a much faster rate than Excel files. CSV Files also does not manipulate data and stores as it is.

Let us consider that an excel file named as 'plant' which stores three variables, named as ph, pd and yield is created inside the folder named as 'My Data' which is placed in the desktop. Thus the path of the file is "C:\\Desktop\\My Data\\plant"

To import excel file the package will be readxl and the command will be library(readxl)

The format of the command to import the excel file is:

```
library(readxl)

dataset <- read_excel(NULL)

View(dataset)
```

Here data set is the name which we want to give for the data set file in R. The name of the data set may or may not be same as the name of original excel file. Inside the bracket in place of NULL we would write the name of the excel file along with its path. Here the name of the file along with its path is C:\\Desktop\\My Data\\plant.xlsx. Thus the command for importing the file will be

library(readxl)

planting <- read_excel("C:\\Desktop\\My Data\\plant.xlsx")

View(planting)

The package for importing csv file is readr.

The format of the command to import the csv file is:

library(readr)

dataset <- read_csv(NULL)

View(dataset)

Here data set is the name which we want to give for the data set file in R. The name of the data set may or may not be same as the name of original excel file. Inside the bracket in place of NULL we would write the name of the excel file along with its path. Here the name of the file along with its path is C:\\Desktop\\My Data\\plant1.csv. Thus the command for importing the file will be

library(readr)

planting1 <- read_excel("C:\\Desktop\\My Data\\plant1.csv")

View(planting1)

(Please note that you have to be careful in upper case and lower case letters. In the function View() the 'V' should always be in upper case).

library(readxl)

Plant_data <- read_excel("C:/Users /Desktop/Rdata/plant1.xls")

View(Plant_data)

Data file imported from R environment. A file can be imported to R session in excel or csv or text format by using the import file menu in the environment window in following manner:

- Click on 'import data set' option under 'environment' menu in the RStudio window.
- Select the appropriate option to select the required file (See Figure 7). If we want to import excel file then click on From Excel and if we want to import a csv file then click on From text (readr).
- If we click on 'From Excel' in the pop up menu option then the import Excel File window will be opened.

- In the browse column of the opened window (see figure 8) the file to be imported is selected from its path. If first row contains names of variables then we have to check the box named as first row as names. If the excel file contains more than one sheet then we have to select the sheet contains our required data.

Fig. 7: Importing file in environment window

- If we click on 'From Text(readr)' in the popup menu option then the Import Text Data window will be opened.
- In the browse column of the opened window (see figure 9) the file to be imported is selected from its path. If first row contains names of variables then we have to check the box named as first row as names.

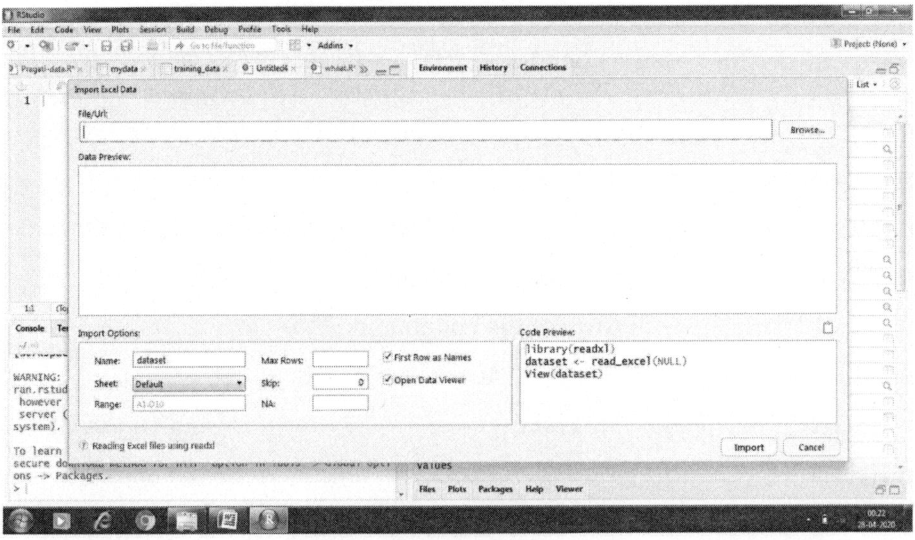

Fig. 8: Import Excel Data window

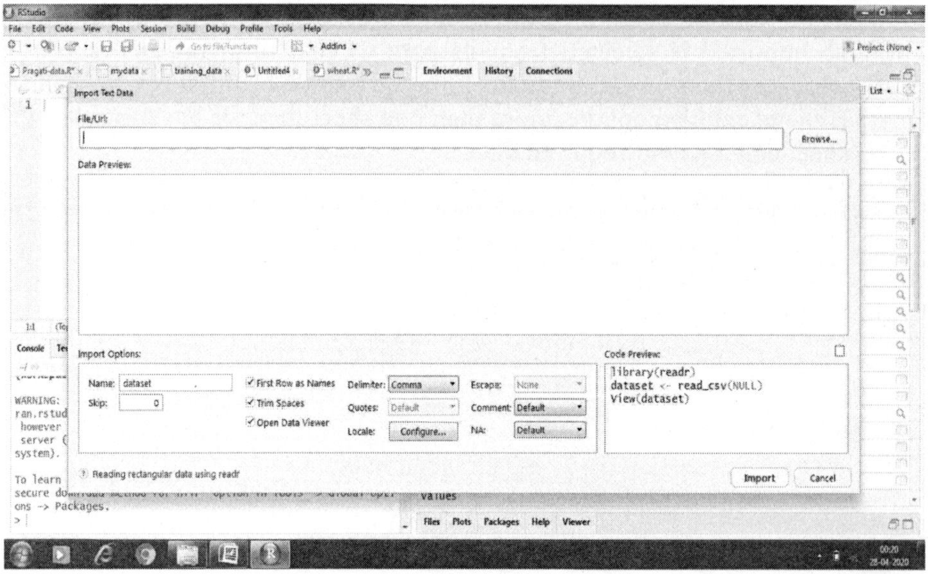

Fig. 9: Import Text Data window

We can also import file by using file.choose() inside the read.csv() function for importing csv file or read.xls() function for excel file. If the argument header is set TRUE, then the first row of the imported data file will be considered as a heading and not get included in the analysis.

Let us create a file in excel which will be having 6 columns and 10 rows, the first row containing name of the variables. Let the name of the 6 column headings be sn, ph, pp, pd, ngl, yld,

w1=read.csv(file.choose(),header=TRUE) # *We can select the data file which in our case is plant2.csv. The csv file named plant2 will be stored in R in name of w1.*

w1 # *will show contents of the file*

names(w1)# *will show the name of variables i.e column headings in the file w1*

head (w1) # *will show first 6 rows of the file w1*

w1[,2:4] # will show all rows of the column no. 2 to 4

w1[3,] # *will show all columns of the row no. 3*

w1[3:7,2:3] # *will show the contents of 3rd to 7th row of 2nd to 3rd column*

w1[c(2,3),c(2,4)]# *will show the contents of 2nd and 3rd row of 2nd and 4th column*

w1[,c(2,4)] # *will show the contents of all rows of 2nd and 4th column* w1[c(2,4,6,9),] # *will show the contents of all columns of 2nd, 4th, 6th and 9th row.*

View(w1)# *to view the data in another Rscript .*

attach(w1) # *If file is attached then the variables in the file can be directly used, otherwise to use a variable we have to write the name of the file followed by the name of the variable with a '$' between them.*

The use of attach function() will be explained in forthcoming chapters.

Getting and setting the working directory in R:

The command setwd() is used to set a working directory. The working directory is set along with the path of the directory. The directory named Plant Data is set as the working directory which is placed in desktop.The format of the command is as follows:

setwd("C:\\ Desktop\\ Plant Data ")

The command getwd() is used to get the working directory. Now the working directory named as Plant Data will contain all the files created in the current session of R. The format of the command is:

getwd()

6

Matrix operations in R

Matrix is a rectangular array of numbers arranged in rows and columns. The order of a matrix is written as r x c, where r is the no. of rows and c is the no. of columns. The numbers included in a matrix are called elements of the matrix. The name of the matrix is denoted by an uppercase English alphabet.

Examples of matrices: $A=\begin{bmatrix}4 & -2 & 1\\ 3 & 0 & 4\\ 2 & 1 & 5\end{bmatrix}_{3X3}$ $B=\begin{bmatrix}3 & -1 & 0\\ 4 & 2 & 6\end{bmatrix}_{2X3}$ $C=\begin{bmatrix}1 & 2\\ -3 & 4\\ 1 & 5\end{bmatrix}_{3X2}$

Here A is a 3 x 3 matrix, B is 2 x 3 matrix, C is a 3 x 2 matrix

To form these matrices in R we take the help of matrix function.

Illustration of use of matrix function:

A<-matrix(c(4,-2,1,3,0,4,2,1,5),nrow=3,ncol=3,byrow=TRUE)

A

The output will be:

	[,1]	*[,2]*	*[,3]*
[1,]	*4*	*-2*	*1*
[2,]	*3*	*0*	*4*
[3,]	*2*	*1*	*5*

The argument by row =TRUE will arrange the elements row wise. The argument nrow takes the value of no. of rows and ncol takes the value of no. of columns.

If byrow = FALSE then the elements will be arranged columnwise.

Illustration of making byrow=FALSE

A1<-matrix(c(4,-2,1,3,0,4,2,1,5),nrow=3,ncol=3,byrow=FALSE)

A1

The output will be:

	[,1]	[,2]	[,3]
[1,]	4	3	2
[2,]	-2	0	1
[3,]	1	4	5

The matrix A1 obtained by making byrow=FALSE will be the transpose of the matrix A obtained by making byrow =TRUE. i.e. rows and columns of matrix A has been exchanged to get matrix A1.

$$A1 = \begin{bmatrix} 4 & 3 & 2 \\ -2 & 0 & 1 \\ 1 & 4 & 5 \end{bmatrix} \text{; Here, } A1 = A'$$

To Find sum of columns and rows the function colSums() and rowSums() are used, where the name of the matrix is written inside the bracket.

Illustration of colSums() and rowSums() to find sum of elements of rows and columns of the matrix A given below

$$A = \begin{bmatrix} 4 & -2 & 1 \\ 3 & 0 & 4 \\ 2 & 1 & 5 \end{bmatrix}_{3\times 3}$$

A=matrix(c(4,-2,1,3,0,4,2,1,5),nrow=3,ncol=3,byrow=TRUE)

sc=colSums(A) # *sc is the name of the data object which stores the value of the sum of the columns*

sc

The output will be:

[1] 9 -1 10

sr=rowSums(A) # *sr is the name of the data object which stores the value of the sum of the rows*

sr

The output will be:

[1] 3 7 8

To Find mean of columns and rows. the function colMeans() and rowMeans() are used

mc = colMeans(A) *#mc is the name of the data object which stores the value of the mean of the columns*

mc

[1] 3.0000000 -0.3333333 3.3333333

mr = rowMeans(A) *sc is the name of the data object which stores the value of the mean of the rows.*

mr

[1] 1.000000 2.333333 2.666667

Finding the no. of rows and no. of columns of a matrix: The functions nrow() and ncol() are used to find the no. of rows and no. of columns of a matrix where the name of matrix is written inside the bracket.

Illustration to find no. of rows and columns

K=matrix(c(2,3,5,1,7,2),3,2)

K

The output will be:

	[,1]	[,2]
[1,]	2	1
[2,]	3	7
[3,]	5	2

r=nrow(K) *# the function nrow(K) will find no. of rows in matrix K.*

cat("no. of rows in the matrix K =", r) *#cat() function is concatenation function which is used to combine two or more outputs*

The output will be:

no. of rows in the matrix K = 3

c=ncol(K)

cat("no. of rows in the matrix K =", c)

The output will be:

no. of columns in the matrix K = 2

The dim() function will give the no. of rows and no. of columns.

Illustration of dim() function:

```
v=dim(K)
v
```

The output will be:

[1] 3 2

The first digit of the output of dim() function which is 3 shows the no. of rows and the second digit which is 2 shows the no. of columns in the matrix K.

Transpose of a matrix: Transpose of a matrix involves the interchanging of rows and columns. To find the transpose of a matrix, the function used in R is t(), where name of the matrix is written inside the bracket.

Illustration of finding transpose of a matrix in R:

```
A = matrix(c(4,-2,1,3,0,4,2,1,5),nrow=3,ncol=3,byrow=TRUE)
B =t(A)
B
```

The output will be:

	[,1]	[,2]	[,3]
[1,]	4	-2	1
[2,]	3	0	4
[3,]	2	1	5

The matrix B is transpose of the matrix A i.e. B = A'

Diagonal matrix: In R, to prepare a diagonal matrix with diagonal elements 2,-1,4,5 the function diag() is used where the name of the data object containing the elements to be placed as diagonal elements is written inside the bracket. The diagonal matrix whose diagonal elements are 2,-1,4,5 is represented by

$$C=\begin{bmatrix} 2 & 0 & 0 & 0 \\ 0 & -1 & 0 & 0 \\ 0 & 0 & 4 & 0 \\ 0 & 0 & 0 & 5 \end{bmatrix}$$

Illustration of forming a diagonal matrix in R:

```
X=c(2,-1,4,5)
Y=diag(X) #The function diag(X) will form the diagonal matrix with the given elements in data object X as the diagonal elements.
```

Y

The output will be:

	[,1]	*[,2]*	*[,3]*	*[,4]*
[1,]	*2*	*0*	*0*	*0*
[2,]	*0*	*-1*	*0*	*0*
[3,]	*0*	*0*	*4*	*0*
[4,]	*0*	*0*	*0*	*5*

The diag() function can also be used to display the diagonal elements of a diagonal matrix.

Illustration to display diagonal elements of a diagonal matrix:

diag(Y)

The output will be:

[1] 2 -1 4 5

Unit matrix: The diagonal matrix in which all the diagonal elements are equal to 1 is known as unit matrix.

Unit matrix of order 3, $I_{3 \times 3} = \begin{bmatrix} 1 & 0 & 0 \\ 0 & 1 & 0 \\ 0 & 0 & 1 \end{bmatrix}$

Illustration of forming a unit matrix:

Diag(3) # *The number inside the matrix denotes the order of the unit matrix.*

The output will be:

	[,1]	[,2]	[,3]
[1,]	1	0	0
[2,]	0	1	0
[3,]	0	0	1

Matrix operations

(i) Addition of matrices

Two or more matrices can be added only when they are of the same order.

$$A = \begin{bmatrix} 2 & 0 & -1 \\ -1 & 3 & 4 \end{bmatrix}; \ B = \begin{bmatrix} 3 & -5 & 8 \\ -2 & 1 & 4 \end{bmatrix}; \ A+B = \begin{bmatrix} 5 & -5 & 7 \\ -3 & 4 & 8 \end{bmatrix}$$

Illustration of addition of matrices in R:

A=matrix(c(2,-1,0,3,-1,4),2,3,byrow=FALSE)

A

The output will be:

	[,1]	[,2]	[,3]
[1,]	2	0	-1
[2,]	-1	3	4

B=matrix(c(3,-5,8,-2,1,4),2,3,byrow=TRUE)

B

The output will be:

	[,1]	[,2]	[,3]
[1,]	3 -	5	8
[2,]	-2	1	4

A+B

The output will be:

	[,1]	[,2]	[,3]
[1,]	5 -	5	7
[2,]	-3	4	8

Illustration of Scalar multiplication of matrices:

Z=5*A

Z

The output will be:

	[,1]	*[,2]*	*[,3]*
[1,]	*10*	*0*	*-5*
[2,]	*-5*	*15*	*20*

K= 6*A-3*B

The output will be:

	[,1]	*[,2]*	*[,3]*
[1,]	*3*	*15*	*-30*
[2,]	*0*	*15*	*12*

Matrix Multiplication: Two matrices can be multiplied if the no.of columns of the first matrix is equal to no. of rows of second matrix. The no. of rows in the resultant matrix is same as no. of rows in the first matrix and the no. of columns is same as the no. of columns in the second matrix. Matrix multiplication is not commutative.

Illustration of Matrix multiplication

A=matrix(c(2,-1,0,3,-1,4), 3,2byrow=FALSE)

B=matrix(c(3,-5,8,-2,1,4),2,3,byrow=TRUE)

D1=A%*%B #multiplication of matrix A with B

D1

The output will be:

	[,1]	*[,2]*	*[,3]*
[1,]	*21*	*19*	*9*
[2,]	*39*	*41*	*19*
[3,]	*51*	*59*	*27*

D2=B%*%A #multiplication of matrix B with A

D2

The output will be:

	[,1]	*[,2]*
[1,]	*39*	*62*
[2,]	*28*	*50*

Finding the determinant of a matrix

To find determinant of a square matrix, the function det() is used where the name of the data object containing the matrix is written.

Let us consider a matrix A=$\begin{bmatrix} 2 & 4 \\ 3 & 5 \end{bmatrix}$; det(A) or $|A|$ =(2 x 5)-(4 x 3) =10-12=-2

Illustration:

A=matrix(c(2,3,4,5),2,2)

A

The output will be:

	[,1]	*[,2]*
[1,]	*2*	*4*
[2,]	*3*	*5*

det(A)# find the determinant of matrix A

The output will be:

[1] -2

We can also print the result as determinant = -2 by using the cat() function. cat() function is concatenation function which is used to combine two or more outputs.

Illustration of cat() function:

A=matrix(c(2,3,4,5),2,2)

B=det(A)

C=cat("determinant =",B)

C

The output will be:

determinant = -2

To find inverse of a matrix: The function solve() is used to find inverse of the square matrix whose name is written inside the bracket.

Illustration of finding inverse of a square matrix:

A<-matrix(c(4,-2,1,3,0,4,2,1,5),3,3)

A

The output will be:

	[,1]	*[,2]*	*[,3]*
[1,]	*4*	*3*	*2*
[2,]	*-2*	*0*	*1*
[3,]	*1*	*4*	*5*

solve(A) *# to find the inverse of matrix A.*

The output will be:

	[,1]	*[,2]*	*[,3]*
[1,]	*-4*	*-7*	*3*
[2,]	*11*	*18*	*-8*
[3,]	*-8*	*-13*	*6*

To solve a system of linear equations by use of matrix algebra

Let us consider a system of 3 linear equations as follows:

$Y_1 = A_1X_{11} + A_2X_{12} - A_3X_{13}$; $Y_2 = A_1 X_{21} + A_2 X_{22} - A_3 X_{23}$; $Y_3 = A_1 X_{31} - A_2 X_{32} - A_3 X_{33}$

The equations can be written in the form of Y = AX,

$$\text{Where, } Y = \begin{bmatrix} Y_1 \\ Y_2 \\ Y_3 \end{bmatrix}_{3X1} ; X = \begin{bmatrix} X_{11} & X_{12} & X_{13} \\ X_{21} & X_{22} & X_{23} \\ X_{31} & X_{32} & X_{33} \end{bmatrix}_{3X3} ; A = \begin{bmatrix} A_1 \\ A_2 \\ A_3 \end{bmatrix}_{3X1}$$

The solution of matrix A is given by $A = X^{-1}Y$. Here A will be a 3 x 1 matrix. The three elements of the matrix A will give the solution for A_1, A_2 and A_3

Illustration:

Y=matrix(c(29,21,7,9),4,1)

Y

The output will be:

	[,1]
[1,]	*29*
[2,]	*21*
[3,]	*7*
[4,]	*9*

X=matrix(c(5,6,-3,8,3,-2,5,-2,-4,1,3,-2,4,6,-1,3),4,4,byrow=TRUE)

X

The output will be:

	[,1]	[,2]	[,3]	[,4]
[1,]	5	6	-3	8
[2,]	3	-2	5	-2
[3,]	-4	1	3	-2
[4,]	4	6	-1	3

solve(X,Y) *##he function solve(X,Y) will find the product of inverse of matrix X with Y*

The output will be:

	[,1]
[1,]	-0.1980676
[2,]	-0.6135266
[3,]	6.7729469
[4,]	6.7487923

Thus the solution for the given system of equations will be A1 = -0.1980676, A2 = -0.6135266, A3 = 6.7729469, A4 = 6.7487923

Concatenation of matrices: It is combining of columns/ rows of two or more matrices

Concatenation of matrices can be (i) horizontal (ii) vertical

Horizontal concatenation is the combining of columns of two or more matrices. The function cbind() is used for horizontal concatenation.

Illustration of horizontal and vertical concatenation

A=matrix(c(2,3,4,5),2,2)

B=matrix(c(1,1,7,8),2,2)

A

The output will be:

	[,1]	[,2]
[1,]	2	4
[2,]	3	5

B

The output will be:

	[,1]	[,2]
[1,]	9	7
[2,]	1	8

C=cbind(A,B) #vertical concatenation

C

The output will be:

	[,1]	[,2]	[,3]	[,4]
[1,]	2	4	9	7
[2,]	3	5	1	8

D=rbind(A,B)

D

	[,1]	[,2]
[1,]	2	4
[2,]	3	5
[3,]	9	7
[4,]	1	8

7

Construction of Frequency Distribution by Using R

Frequency Distribution is the arrangement of various values of a variable in order of magnitude either individually or in groups alongwith their corresponding frequencies or class frequencies. Frequency distribution is necessary for data condensation and revealing pattern within the data. Frequency distribution can be (i) Ungrouped Frequency Distribution (ii) Grouped frequency distribution.

In case of ungrouped frequency distribution, the individual values of the variable are arranged in ascending order alongwith their frequencies. Frequency is the number of times a particular observation is repeated.

In case of grouped frequency distribution, the values of variable are grouped into classes alongwith their corresponding class frequencies. Class frequency of a class is the number of observations lying within the upper class limit (maximum value) and lower class limit (minimum value) of the class.

Let us first illustrate the preparation of ungrouped frequency distribution. Consider 50 petridishes each containing 6 no.s of seeds for germination. Let the variable x_i denote the no. of seeds germinated in a petridish.

The no. of seeds (out of 6 seeds) that germinate in a petridish are: 4,2,1,6,3,5,2 ,4,4,1,0,2,1,4,5,6,0,1,4,4,1,3,2,4,5,6,0,1,2,4,4,3,4,5,3,5,4,4,1,0,4,2,2,6,1,4,2,6. These data are entered in 1st column of sheet1 of excel with seeds as the name of the column and the file is named as 'unfq'. This file will be imported to R for construction of ungrouped frequency distribution in R.

To import the excel file named as unfq to R, we have to click on Import Dataset then browse the file alongwith its path.

After importing the file, it is stored in R as a data object. Then the frequency() function is used to find out the frequencies of the individual observations, where the name of the data object is written within the parentheses. Before using the function the data object must be attached to R by using attach() function in which the name of the data object to be attached is written inside the parentheses.

un_freq=unfq# un_freq is the name of data object assigned to the imported excel file unfq.

View(un_freq) # View() function is used to view the contents of the file.

Thus the R codes for constructing the ungrouped frequency distribution in R are as follows:

attach(un_freq)

frequency=table(seeds)

frequency

The output will be:

seeds

seeds

0 1 2 3 4 5 6

4 8 8 4 14 5 5

cbind(frequency)

The output will be

```
  frequency
0    4
1    8
2    8
3    4
4    14
5    5
6    5
```

Now let us illustrate the preparation of grouped frequency distribution. Consider the yield of a crop in 30 plots of a farm. Let x_i be the yield (in Quintals/plot of the plots). The values of x_i are 22.3, 34.2, 44.2, 25.0, 41.5, 21.4, 33.4, 40.0, 48.4, 54.3, 50.0, 52.3, 42.4, 41.2, 30.0, 31.2, 24.2, 28.4, 34.4, 41.5, 44.3, 31.2, 24.6, 27.5, 50.2, 44.5, 48.5, 43.2, 49.0, 35.0

These data are entered in 1st column of sheet1 of excel with yield as the name of the column and the file is named as 'grfq'. This file will be imported to R for construction of grouped frequency distribution in R.

To import the excel file named as grfq to R, we have to click on Import Dataset then browse the file alongwith its path. After importing the file, it is stored in R as a data object which we have named here as 'grouped_FD'. Then the data object must be attached to R by using attach() function in which the name of the data object to be attached is written inside the parentheses.

We can also use the following code to import an excel file

library(readxl)

grouped_FD <- read_excel("C:/Users/PT/Desktop/My Data for R/grfq.xls")

The path mentioned inside the parentheses may vary from as depending upon the system and the user.

grouped_FD is the name of the data object in R for the imported excel file grfg

View(grouped_FD)

attach (grouped_FD) # *attach() function will attach the data object grouped_FD in R*

range(Yield) # *range() will find the minimum and maximum value of variable 'Yield' in the data object grouped_FD*

The output will be:

[1] 21.4 54.3

limits=seq(20,55,by=5) # *seq() function will set the class limits.*

limits

The output will be:

[1] 20 25 30 35 40 45 50 55

In seq() function, the lowest and highest values of the sequence are mentioned by the user and the interval is mentioned after by=

intervals=cut(Yield,limits,right=FALSE) # *right=FALSE will make the classes closed at left and open at right intervals*

The output will be:

[1] [20,25) [30,35) [40,45) [25,30) [40,45) [20,25) [30,35)

[8] [40,45) [45,50) [50,55) [50,55) [50,55) [40,45) [40,45)

[15] [30,35) [30,35) [20,25) [25,30) [30,35) [40,45) [40,45)

[22] [30,35) [20,25) [25,30) [50,55) [40,45) [45,50) [40,45)

[29] [45,50) [35,40)

7 Levels: [20,25) [25,30) [30,35) [35,40) ... [50,55)

frequency=table intervals# table() function calculates frequency of each class

frequency

The output will be:

intervals

[20,25) [25,30) [30,35) [35,40) [40,45) [45,50) [50,55)

4 3 6 1 9 3 4

cbind(frequency)# cbind() function will make the result to appear in column form

The output will be:

frequency

[20,25) 4

[25,30) 3

[30,35) 6

[35,40) 1

[40,45) 9

[45,50) 3

[50,55) 4

8

Diagrammatic and Graphical Presentation in R

Diagrams and graphs are the effective and attractive ways of presenting data. Widely used diagrams for data presentation are pi chart and bar diagrams. Bar diagrams are also of various types; (i) Simple bar diagram (ii) Component or sub-divided bar diagram (iii) Multiple bar diagram (iv) Percentage bar diagram. The widely used graphical presentation are histograms, frequency polygon and frequency curve.

There are numerous libraries to create charts and graphs in R programming language.

Pie-chart- A pie-chart is a representation of values of variable as sectors of a circle with different colours/shadings. The colours/shadings used for different sectors are indexed at the side of the graph. **pie()** function is used to create pie chrt in R.

Let us create an exel file names as 'exp_sports' which contains the expenditure made (in Rs.) under various sports in a sports club. The name of the sports are cricket, hockey, tennis and football. The names of sports are placed under the column head 'sports'. The corresponding expenditure (in Rs.) are 7200, 3600, 1800, 1800 and they are placed under the column head 'expenditure'.

The basic syntax for creating a pie-chart using the R is –

```
pie(y, labels, radius, main, col, clockwise)
```

Following is the description of the parameters used –

- **y** is a vector containing the numeric values used in the pie chart. The vector y denotes the values under the column head 'expenditure' in the data object named as 'pc'. Since in our program we will use the attach(pc) function.
- **labels** is used to give description to the sectors. The labels parameter are the name of the sports uner the column head 'sports'.

- **radius** indicates the radius of the circle of the pie chart.(value between −1 and +1).
- **main** indicates the title of the chart. In our case the title of the chart will be given 'Expenditure in sports by sports club'
- **clockwise** is a logical value indicating if the sectors are drawn clockwise or anti clockwise.

Illustration of using pie() function to create pie chart:

First we will import the excel file named as Pie_chart (the contents of the file is given in Annexure 1). which contain the expenditure data under different sports in a sports club. After importing the excel file, let us store the file in R as a data object under the name 'pc' and then attach the data object sports by using attach() function so that the variables (which are the column heads) of the data object can be used directly in the program which mentioning the data object.

The R program for creating pie chart is as follows:

pc=Pie_charts

attach(pc)

pie(expenditure,sports,radius=2,main="Expenditure for sports by sports club", clockwise=FALSE)

The output will be:

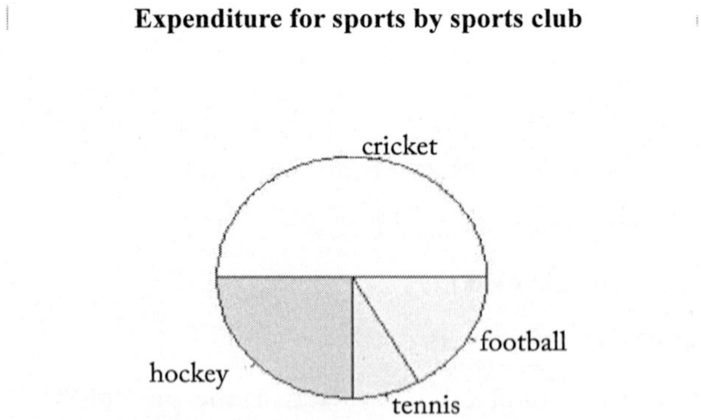

Construction of simple bar diagram: A bar diagram represents data in rectangular bars whose length is proportional to the value of the variable. The function barplot() is used to create bar charts in R. We can draw both vertical and horizontal bars in R. In bar diagram each of the bars can be given different colours.

Let us create an excel file named as 'Bar1' which contains the profit (in Rs. lakh) made by a firm in the years 2016-17, 2017-18. 2018-19 and 2019-20. The years are stored under the column head named as 'Year' and the corresponding profit are stored by the column head named as 'Profit'. The excel file 'Bar1'(the contents of the file is given in Annexure 2). is imported to R and then saved in R as data object and saved by the name 'SBD'. The data object SBD is then attached by using attach(SBD) function so that the variables Year and Profit in the data object SBD can be directly accessed.

The syntax to create a bar-chart in R is –

barplot(H,xlab,ylab,main, names.arg,col)

Following is the description of the parameters used –

- H is a vector or matrix containing numeric values used in bar chart. Here we write the name of column head (i.e. variable) containing the profit values which in our case is 'Profit'.
- xlab is the label for x axis. Here we write the name of the x-axis. In our case it will be 'Year'
- ylab is the label for y axis. Here we write the name of the y-axis. In our case it will be 'Profit (in Rs. Lakh)'
- main is the title of the bar chart. Here the title which we give to the bar diagram is 'Simple bar diagram'
- names.arg is a vector of names appearing under each bar.

col is used to give colours to the bars in the graph.

The R program for constructing simple bar diagram after importing the excel file named as 'Bar1' (which contains the yearwise data of profit (in Rs. Lakh) made by a firm) will be as follows:

SBD=Bar1

attach(SBD)

View(SBD)

barplot(ProfitAxlab="Year",ylab="Profit", main="Simple bar diagram",col = "blue", names.arg=Year)

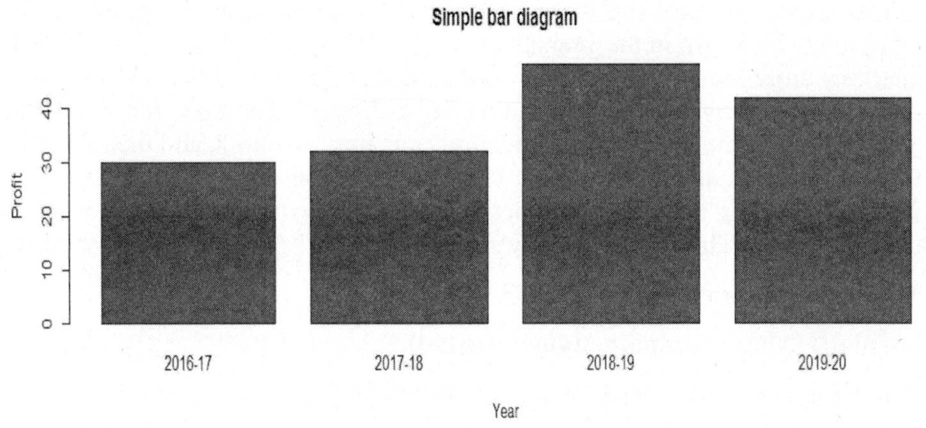

The above bar diagram can also be constructed by directly creating the data in the program and for this we do not need to import any file. An illustration of this has been given below:

#creating data

D=c(30,32,48,42)

K=c("2016-17", "2017-18", "2018-19", "2019-20")

To plot the bar diagram

barplot(D,xlab= "Year",ylab= "Profit (in Rs. lakh)", main= "Simple bar diagram", names.arg=K)

The output will be same as in the above program:

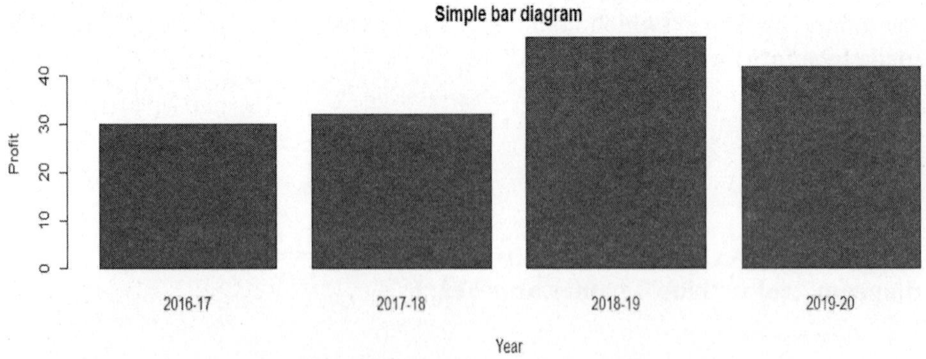

Bar diagram for categorical data: The count for each item can be plotted as bar diagram by using categorical data. Let us consider the data for category of farmers.

The R program for construction of bar plot for categorical data is:

category_farmers=c("Small","Large"," Medium", "Marginal", "Small", "Large", "Marginal", "Small")

table(category_farmers)

The output will be:

category_farmers

Large Marginal Medium Small

3 2 2 4

barplot(table(category_farmers),density=20,main="category of farmers", xlab="Category of Farmers",ylab="No.of farmers",col="green")

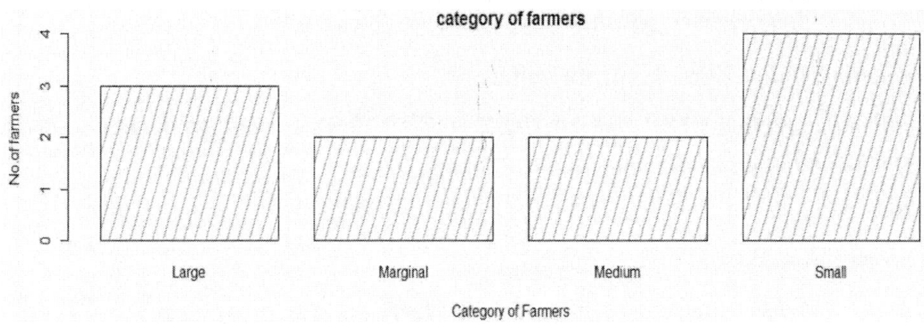

Sub-divided or component or stacked bar diagram

In this type of bar diagram, the bars are sub-divided into different components. Each subdivision of the bars represents the component value of the character studied. This is used when each individual data can be sub-divided into its own components. Here each component of individual data are compared and also the individual data for each group can be compared. For e.g., let us consider the yearwise profit (in Rs. lakh) made by 3 firms named as A,B,C. Here year wise total profit made by 3 firms is considered as individual data and the profit made by a firm in a particular year is considered as a component of the total profit made by all the 3 firms in that year. Here we can compare the profit made by each firm in a year.

We will construct the subdivided bar diagram by creating data in the program itself:

Let us consider the following data in tabular form for our subdivided bar diagram.

Year	Profit (in Rs. Lakhs) by Firm A	Profit (in Rs. Lakhs) by Firm B	Profit (in Rs. Lakhs) by Firm C
2016-17	12	18	10
2017-18	16	24	20
2018-19	12	16	7
2019-20	20	12	18

Here the data consists of yearwise profit (in Rs. lakhs). Each yearwise profit is subdivided into year wise profit made by three firms. Thus R program to create the sub-divided bar diagram will be as follows:

#creation of input vectors

firms=c("A","B","C")

colours=c("red","blue","green")

years=c("2016-17","2017-18","2018-19","2019-20")

#creation of matrix of the profit values

profit=matrix(c(12,18,10,16,24,20,12,16,7,20,12,18), ncol = 4, nrow = 3, byrow = TRUE)

profit

The output will be:

```
      [,1]   [,2]   [,3]   [,4]
[1,]  12     18     10     16
[2,]  24     20     12     16
[3,]  7      20     12     18
```

#creation of the sub-divided bar diagram

barplot(profit,names.arg=years,xlab="Years",ylab="profit (in Rs. lakh)",main="Sub-divided bar diagram",col=colours)

#creation of legend of the diagram

legend("topright",cex=.5,firms,fill = colours)

The output will be:

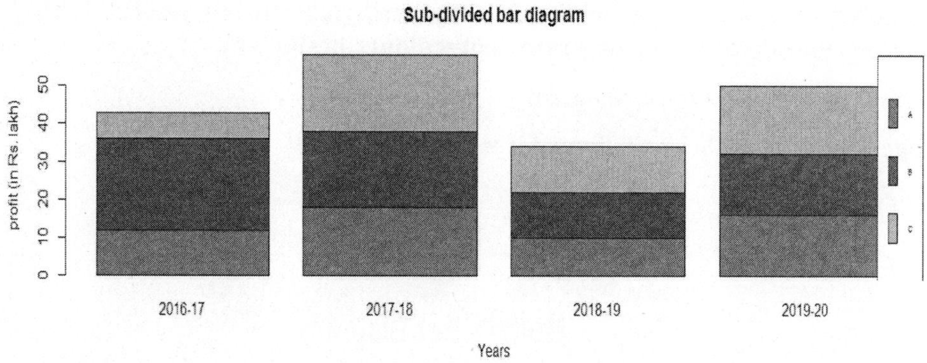

Multiple bar diagram: In some bar diagrams, the groups are divided into two or more categories and the values of each category of a group are represented by different bars which are placed side by side without any gap and space is left between the bars of different groups. Such type of diagrams are called multiple bar diagrams. Here multiple bars, placed side by side without any gap are present within a group. In our above example, Year is regarded as groups. We can create multiple bar diagram for the above data by using the following program as illustrated below:

#creation of input vectors

firms=c("A","B","C")

colours=c("red","blue","green")

years=c("2016-17","2017-18","2018-19","2019-20")

#creation of matrix of the profit values

profit = matrix (c(12,18,10,16,24,20,12,16,7,20,12,18), ncol = 4, nrow = 3, byrow = TRUE)

profit

The output will be:

	[,1]	*[,2]*	*[,3]*	*[,4]*
[1,]	*12*	*16*	*12*	*20*
[2,]	*18*	*24*	*16*	*12*
[3,]	*10*	*20*	*7*	*18*

#creation of the multiple bar diagram

barplot(profit,names.arg=years,xlab="years",ylab="profit (in Rs. lakh)", main="Sub-divided bar diagram",col=colours,beside=T)

#creation of legend of the diagram

legend("topright",cex=.5,firms,fill = colours)

The output will be:

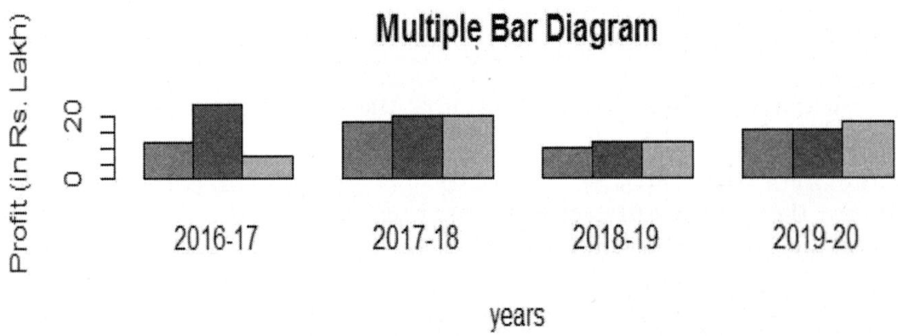

Construction of histogram: Histograms are rectangular graphical figures whose x-coordinate shows the class interval and y-coordinate shows the frequency of the class interval.

We can construct histogram in R by using hist() function. The syntax for construction of histogram is hist(x,main,xlab, xlim, ylim, breaks, col, border)

Description of the parameters used in the function hist():

x is the vector containing the values used in histogram.

main is used to provide title of the graph.

xlab is used to give description to x-axis.

xlim is used to specify the range of values on the x-axis

ylim is used to specify the range of values on the y-axis.

breaks is used to mention the width of each bar.

col is used to set colours to the bars.

border is used to set the border colour of each bar.

Illustration:

#creation of histogram

yield=c(12.2,13.5,23.4,10.4,15.4,24.5,28.4,30.0, 32.8,27.8,25.0,22.4,13.4,16.3, 32.8)

yield

The output will be:

[1] 12.2 13.5 23.4 10.4 15.4 24.5 28.4 30.0 32.8 27.8 25.0

[12] 22.4 13.4 16.3 32.8

hist(yield)

The output will be:

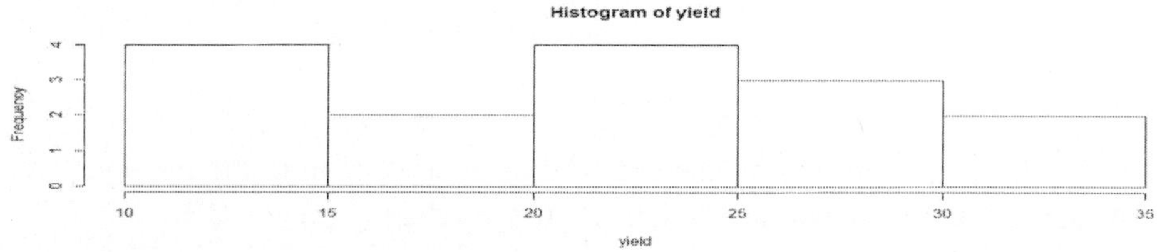

hist(yield,xlab ="yield", xlim=c(10,35), ylim=c(0,5), col="blue", border = "red", breaks = 5)

The output will be:

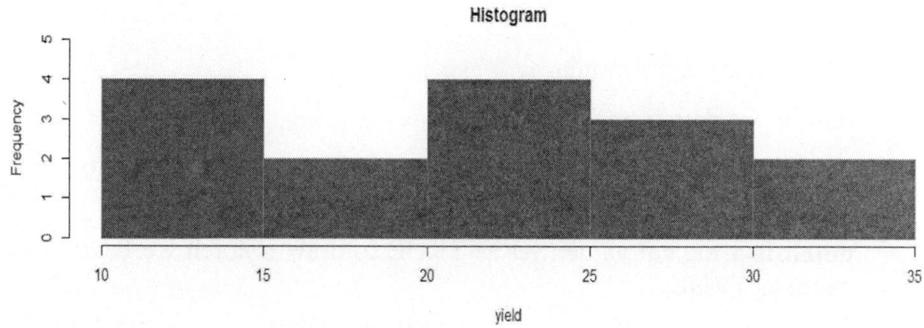

hist(yield,xl lab= "yield", xlim=c(0,35), ylim=c(0,5), col="blue", border ="red", breaks = 5, main="Histogram of yield")*# limit for x –axis is 0 to 35*

Box plots: Box plot shows the distribution of data in a data series. Box plot represents the minimum, maximum, first quartile, second quartile (median) and third quartile of the data set.

The general outlook of a box-plot is:

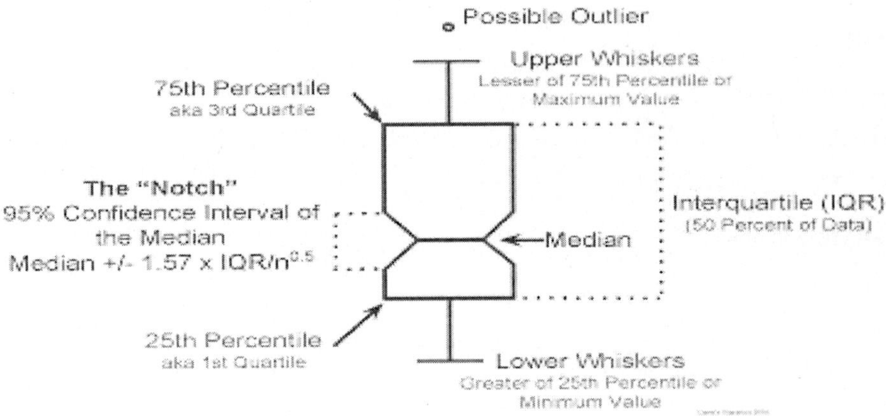

The notch in box-plot shows the confidence interval around the median. This confidence interval is calculated normally as: median $\pm$ 1.57 x IQR/.$\sqrt{n}$

In R, box-plots are created by using boxplot() function.

The syntax for construction of box plot in R is

Boxplot(x, data, varwidth, notch, names, main)

The description of the parameters is:

- **x** is a vector or a formula.
- **data** is the data frame.
- **varwidth** is a logical value. To draw width of the box proportionate to the sample size we have to set varwidth as TRUE.
- **notch** is a logical value. Set as TRUE to draw a notch we have to set notch as TRUE.
- **names** are the group labels which will be printed under each boxplot.
- **main** is used to give a title to the box plot.

Illustration: Let us consider a data on production of two crops, crop A and crop B (in '000tonnes) of odisha for 30 years. Let us make an excel file for the data with the variable named as crop A for production of cereals (in '000tonnes) and crop B for production of pulses (in '000tonnes). Let us name the file as 'Box_ plot'(the contents of the file is given in Annexure 3). The file is to be imported to R by following the procedure described earlier. Then the file is stored in R as a data object with name 'box'. The R program for the construction of box plot is :

box=Box_plot

View(box)

attach(box)

boxplot(box,ylim= c(1000,8000), xlab="crops",ylab="production", notch= TRUE, varwidth = TRUE, col=c("green","blue"),names=c("crop A","crop B"), breaks=500)

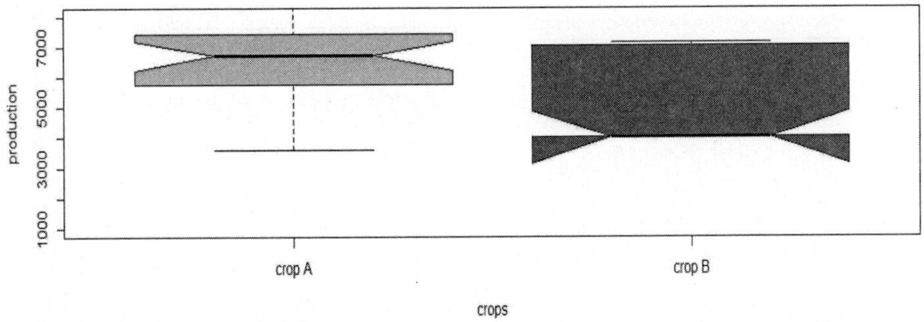

Line graph: It is a graph obtained by joining a series of points consecutively by help of lines. The points are generally ordered in any one axis which is preferably the x-axis. The magnitudes of the points are mentioned in y-axis. These graphs are usually used for studying trend in the data. The function used for construction of line graph in R is plot().

The syntax of plot() function is: plot(x,type,col,xlab,ylab)

The description of the parameters is:

- **x** is a vector containing the numeric values.
- **type** takes the value "p" to draw only the points, "l" to draw only the lines and "o" to draw both points and lines.
- **xlab** is the label for x axis.
- **ylab** is the label for y axis.
- **main** is the title of the graph.
- **col** is used to give colours to both the points and lines.

Illustration: Let us consider a data on production of a crop (in '000tonnes) of odisha for 30 years. Let us make an excel file for the data with the variable named as prod for production of the crop. Let us name the file as 'Line_graph'(the contents of the file is given in Annexure 4). The file is to be imported to R by following the procedure described earlier. Then the file is stored in R as a data object with name 'line'. The R program for the construction of box plot is:

line=Line_graph

View(line)

attach(line)

plot(prod) *#no use of parameters*

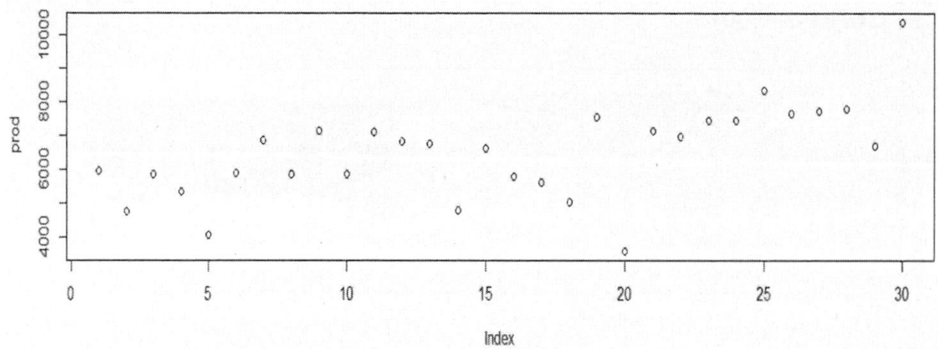

In the above R program if we write

plot(prod,type="l")# *use of type parameter*

The output will be:

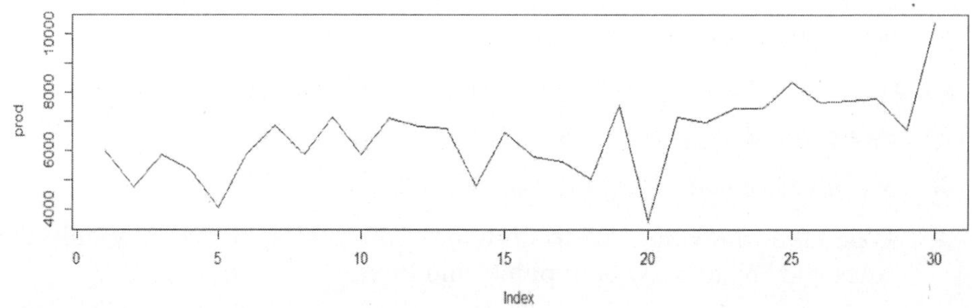

plot(prod,type="o")# *use of type parameter*

The output will be:

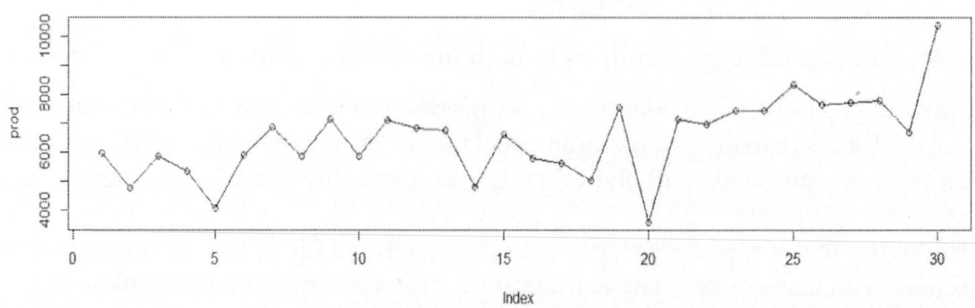

plot(prod,type="p")*#use of type parameter*

The output will be:

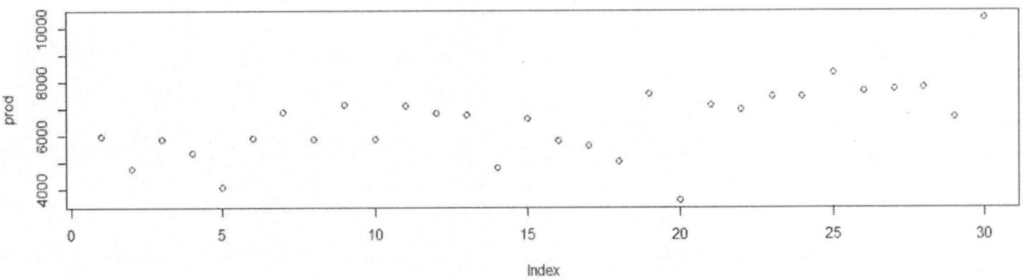

plot(prod,xlab="time",ylab="production",col=c("green","red"),type="o")

The output will be:

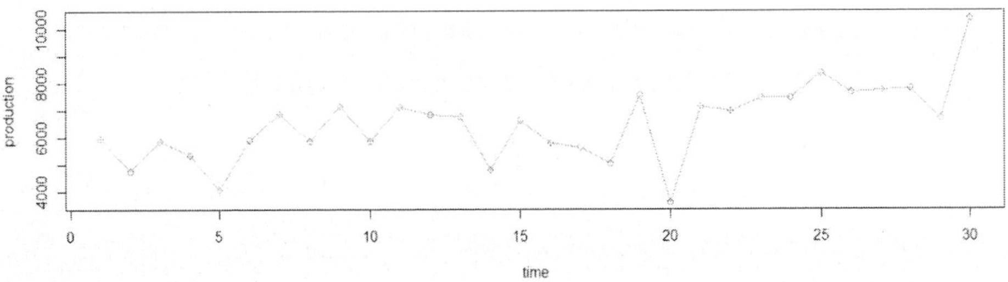

Multiple lines: More than one line can be drawn on the same graph by using the lines() function. After the first line is plotted, the lines() function can use an additional vector as input to draw the second line in the graph. Let us consider a data on production (in '000tonnes) of two crops of odisha for 30 years. Let us make an excel file for the data with the variable named as <u>prod1</u> and <u>prod2</u> for production of the crop. Let us name the file as 'Line_graph'(the contents of the file is given in annexure 5). The file is to be imported to R by following the procedure described earlier. Then the file is stored in R as a data object with name 'line1'.

The R program will be:

plot(prod1, type ="l",col ="red",xlab="time",ylab="production")

lines(prod2,col="blue")

The output will be:

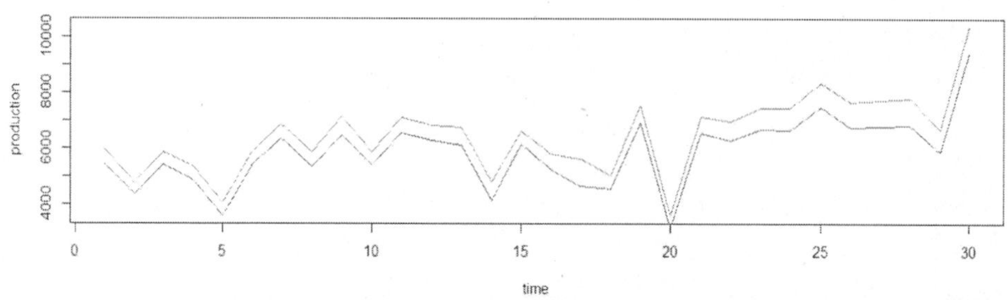

Scatter diagram: It is the diagram of dots plotted in the Cartesian plane. Each dot (or point) represent the value of both the variables plotted in the Cartesian plane. Scateer is also created by using plot() function.

The syntax for using plot() function to construct scatter plot is:

plot(x,y,xlab,ylab,main,xlim,ylim,axes)

- **x** is the data set whose values are the horizontal coordinates.
- **y** is the data set whose values are the vertical coordinates.
- **main** is the tile of the graph.
- **xlab** is the label in the horizontal axis.
- **ylab** is the label in the vertical axis.
- **xlim** is the limits of the values of x used for plotting.
- **ylim** is the limits of the values of y used for plotting.
- **axes** indicates whether both axes should be drawn on the plot.

Let us consider the example of a height (in cm) and weight (in kg) of 30 persons. Let us create an excel file for this data and name the file as 'scatter_ plot' .(The contents of this file is given in Annexure 6)The values of height are given under the variable named as 'ht' and the values of weight are given under the variable named as 'wt'. The file is imported to R in the procedure described earlier. The R program for construction of scatter plot is:

s_plot=scatter_plot # *the imported excel file is stored in R as the data object named as 's_plot*

attach(s_plot)

plot(ht,wt)

The output will be:

plot(ht,wt,xlab="height",ylab="weight",main="Scatter Plot") # *use of labels and title*

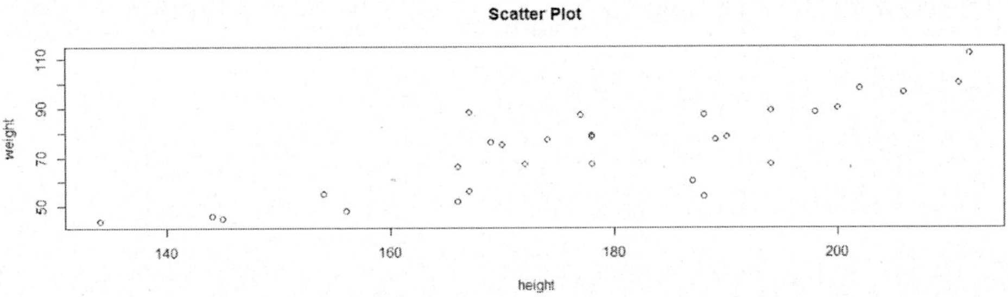
Scatter Plot
weight
50
70
90
110
140
160
180
200
height

9

Descriptive Statistics using R

The branch of statistics which describes the data is called descriptive statistics. After collection, the data are presented in form of tables, diagrams or graphs. This facilitates the further analysis of data. The most basic form of data analysis is descriptive analysis of data collected and presented for the study. The descriptive statistics summarises the data and helps to find the representative value of the entire data set and the variation of the other values in the data set around the representative value. The descriptive statistics includes the measures of central tendency or averages, measures of dispersion or scatteredness, measures of skewness and the measures of kurtosis. Before going to discuss the R codes for various measures of descriptive analysis, we would first discuss in brief about these measure. This will help in better understanding of the R codes.

Measures of central tendency or averages: They are the representative values of the data series. They give an idea about the concentration of the values in the central part of the distribution of the data. There are five types of averages. They are 1. Arithmetic Mean (A.M.) 2. Geometric mean (G.M.) 3. Harmonic Mean (H.M.) 4. Median 5. Mode.

1. Arithmetic mean is the sum of all the observations divided by the number of observation.

   $$\text{A.M. } \left(\bar{x}\right) = \frac{\sum_{i=1}^{n} x_i}{n}$$

   Where n is the number of observations and x_i is the value of the observation.

2. Geometric Mean is the nth root of the product of 'n' no.s of observation.

   $$\bar{G} = (x_1 \times x_2 \times \ldots \times x_n)^{1/n} = \Pi x_i^{1/n}$$

3. Harmonic Mean is the reciprocal of the arithmetic mean of the reciprocal of the given observations.

   $$\bar{H} = \frac{n}{\sum_{i=1}^{n} \frac{1}{x_i}}$$

4. Median:

4. Median – It is the value at the middle position of the data series after arranging the data series in order of magnitude. This value exceeds and is exceeded by same number of observations i.e. it divides the entire data series into two equal parts. To find the median, the given data are arranged either in ascending or descending order of their magnitude. If 'n' is odd, then median = value of $\{(n+1)/2\}^{th}$ observation.

If 'n' is even, then

median = [value of $\frac{n}{2}^{th}$observation +value of $\{(n+1)/_2\}^{th}$ observation]/2

Partition Values

These are the values which divide the series into equal parts

They may be:

i. Quartiles – divides the series into 4 equal parts.
ii. Pentiles - divides the series into 5 equal parts.
iii. Octiles - divides the series into 8 equal parts.
iv. Deciles – divides the series into 10 equal parts.
v. Percentiles – divides the series into 100 equal parts.

Thus there are 3 quartiles, 4 pentiles, 7 octiles9 deciles and 99 percentiles.

5. Mode – It is the value of the variable which has the highest frequency.

If there are more than one mode in a data series, then a single approximate value of mode can be obtained from the empirical relationship:

Mode = 3×Median - 2×Mean.

Measures of dispersion or scatteredness:

Study of dispersion gives an idea about the homogeneity and heterogeneity of the data series. It determines how much the values of the observation vary from their averages. The series in which the values of the observation differ much from their averages is said to be heterogeneous or highly dispersed. The series in which the value of the observation does not differ much from their averages is said to be homogenous or less dispersed.

Measures of dispersion are range, mean absolute deviation, quartile deviation, standard deviation. The square of standard deviation is known as variance. These are absolute measures of dispersion. Coefficient of Variation is the relative measure of dispersion.

1. **Range** is the difference between largest and smallest observation of a series.

   Range = Largest – Smallest

2. **Mean Absolute Deviation (M.A.D)** is the A.M. of the absolute deviation of the observations from their average (mean, median and mode).

   $$\text{M.A.D} = \frac{1}{n}\sum_{i=1}^{n}|x_i - A|;\ \text{for individual series}$$

   Where A = mean, median or mode

3. **Quartile Deviation** is half of the difference between third and first quartile.

   $$\text{Q.D.} = \frac{1}{2}\ (Q_3\text{-}Q_1)$$

   It is also known as semi-inter quartile range.

4. **Standard Deviation** is the positive square root of the A.M of the square of deviations of the observations from their mean. Denoted as σ.

   $$\sigma = +\sqrt{\frac{1}{n}\sum_{i=1}^{n}\left(x_i - \bar{x}\right)^2}$$

Square of S.D. is known as Variance. It is denoted by σ^2

Coefficient of Variation **(C.V)** is the most important relative measure of dispersion.

$$\text{C.V} = \frac{\sigma}{\bar{x}} \times 100$$

C.V isUsed to compare variability in two or more series.

Series having high C.V is said to be more heterogeneous/less consistent/more inconsistent. Series having low C.V is said to be more homogeneous/more consistent/less inconsistent.

There is also a measure of dispersion which measures the extent of variation between the observations of two variables. This measure s called as covariance.

Covariance between two variables x and y is given as,

$$\text{Cov(x,y) or } \sigma_{xy} = \frac{1}{n}\sum_{i=1}^{n}\left(x_i - \bar{x}\right)\left(y_i - \bar{y}\right)$$

Moments:

The rth moment of a variable X about any point A is given by:

$\mu_r' = \frac{1}{n}\sum_{i=1}^{n}(x_i - A)^r$; for individual or simple series

where 'A' is any number. 'r' is the order of the moment.

If $A=\bar{x}$, then it is known as r[th] central moment & is represented as μ_r .

If A is any number other than $\bar{x}$ (mean) then it is known as r [th]raw moment and is denoted as μ_r'

If A = 0, it is called r[th] raw moment about origin.

Skewness means 'lack of symmetry'. Dispersion measures the scatteredness in data, whereas, skewness provides the direction of scatteredness. Symmetry means If there is no skewness in data there is symmetry. In this case the frequency curve of the data is bell shaped or symmetrical and the mean, median and mode of the data series are equal. If the curve extends more towards right, it is called right tailed curve & here the skewness is positive and mean>median>mode. If the curve extends more towards left, it is called left tailed curve & here the skewness is negative and mean<median<mode.

Measure of skewness is provided by $\beta_1=\mu_3^2/\mu_2^3$ but provided more appropriately by $\gamma_1==\mu_3/\mu_2^{3/2}\beta_1$ is called coefficient of skewness.

Kurtosis determines peakedness of the frequency curve.

β_2 & □$_2$ provides measure of Kurtosis.$\beta_2=\mu_4/\mu_2^2$;$\gamma_2=\beta_2$-3

If the curve is more flatter, then it is called platykurticcurve and here$\gamma_2<0;\beta_2<3$.If the curve is more peaked, then it is called leptokurtic curve and here$\gamma_2>0;\beta_2>3$. If the curve is neither peaked nor flat it is called mesokurtic curve. Here $\gamma_2=0;\beta_2=3$

The curve which is both symmetric and mesokurtic is called normal curve.

Now we will study the conduct of descriptive analysis of the data by using R. We would consider a data set stored in an excel file. First we will describe the data set. The data consists of 3 variables which are leaf length (in cm) , plant height (in cm) and straw yield (in kg). The data on leaf length, plant height and straw yield are entered in column A, B, C respectively. Each column have their respective heading as ll (for leaf length), ph (for plant height) and sy (for straw yield). Under each variable there are 30 data points. The name of the excel file is given as 'descriptive'(the contents of the file is given in Annexure 7).

First the excel file named as descriptive is imported to the R by following the procedure described above. To import the excel file named as dctv to R, we have to click on Import Dataset then browse the file alongwith its path. After importing the file, it is stored in R as a data object in the name 'dctv'.

Illustration:

dctv=descriptive # *the imported file named as 'descriptive' is stored in R as data object named as 'dctv'.*

View(dctv) # *The contents of the data object dctv is viewed.*

names(dctv) # *will display the names of the column heads i.e. the variables.*

The output will be:

[1] "ll" "ph" "sy"

head(dctv) # *will display the first six data points of each column.*

The output will be:

ll ph sy

<dbl> <dbl> <dbl>

1 20.6 282. 69.1

2 22.2 238. 92.6

3 21.3 330. 82.2

4 21.0 302. 120.

5 25.1 300. 30.1

6 21.1 237. 54.3

tail(dctv) # *will display the last six data points of each column.*

The output will be:

ll ph sy

<dbl> <dbl> <dbl>

1 25.1 186. 38.1

2 21.4 198. 27.8

3 16.6 191. 32.0

4 20.8 188. 43.7

5 21.2 154. 35.3

6 20.5 128. 21.2

attach(dctv) # *The file/data object will be attached to R*

Arithmetic mean of one variable at a time

mean(ll)*#without use of any data object for the output*

The output will be:

[1] 20.647

k=mean(ph) *# without use of any data object for the output*

k

The output will be:

[1] 217.0102

p=mean(sy) *#printing with name of the measure*

cat("mean =",mean(sy))

The output will be:

mean = 51.8125

mean(c(ll,ph,sy))*# will find the combined mean of the variables*

The output will be:

[1] 96.48992

c(mean(ll),mean(ph),mean(sy)) *# will find the mean of all the three variables at a time*

The output will be:

[1] 20.6470 217.0102 51.8125

The difference between mean(c(ll,ph,sy)) and c(mean(ll),mean(ph),mean(sy)) is that the former will calculate the combined mean of all the three variables named as ll, ph and sy and will provide a single combined mean, whereas, the later will find the mean of all the three variables individually at a time.

cat(mean(ll),mean(ph),mean(sy)) *# Finding mean of all variables at a time*

The output will be:

20.647 217.0102 51.8125

cat("mean of ll=",mean(ll),"mean of ph=",mean(ph),"mean of sy=",mean(sy))

The output will be:

mean of ll= 20.647 mean of ph= 217.0102

mean of sy= 51.8125

Finding the combined mean can be demonstrated in more lucid manner by taking some data points of different data objects in the program itself.

Illustration:

x=c(2,3,4);y=c(1,2,3);z=c(3,6,9)

mean(x);mean(y);mean(z)

The output will be:

[1] 3

[1] 2

[1] 6

mean(c(x,y,z)) *# combined mean*

The output will be:

[1] 3.666667

Geometric mean

There is no direct function in R to find geometric mean. We have to create a function for geometric mean. We have to create a function for geometric mean and then give a name to the function.

gm=function(g) *# function of g is created and named as gm*

{

return(10^(mean(log10(g))))

}

Now we use the function which is named by us as **gm** to find the geometric mean of each individual variables in our data such as ll.ph and sy.

gm(ll)# to show the value of geometric mean of ll

The output will be:

[1] 20.54494

cat("G.M.=",gm(x)) # to print that g.m.= obtained value

G.M.= 20.54494

gm(ph)

The output will be:

[1] 213.2832

cat("G.M.of ll =",gm(ll),"G.M.of ph =",gm(ph),"G.M.of sy =",gm(sy))

The output will be:

G.M.of ll = 20.54494 G.M.of ph = 213.2832 G.M.of sy = 46.60178

gm(c(ll,ph,sy))# Combined GM.

[1] 58.88726

Illustration of combined GM by taking some data points of different data objects in the program itself.

x=c(2,3,4);y=c(1,2,3);z=c(3,6,9) *# assigning of values to variables x,y and z*

gm(x);gm(y);gm(z) *#find geometric mean of x, y and z*

The output will be:

[1] 2.884499

[1] 1.817121

[1] 5.451362

The output will be:

[1] 3.666667

gm(c(x,y,z)) *#find combined mean of x,y and z*

The output will be:

[1] 3.057171

* we can give any name to the function for geometric mean.

Harmonic mean

Like geometric mean, there is also no direct function in R to find harmonic mean. We have to create a function for harmonic mean. We have to create a function for harmonic mean and then give a name to the function.

hm=function(h)

{return(1/(mean(1/h)))}

hm(ll) *# to show the value of harmonic mean of variable ll*

The output will be:

[1] 20.44167

hm(ph) *# to show the value of harmonic mean of variable ph*

The output will be:

[1] 209.6673

hm(sy) *# to show the value of harmonic mean of variable sy*

The output will be:

[1] 42.56131

cat("h.m.=",hm(ll)) *# to print that h.m.= obtained value*

The output will be:

h.m.= 20.44167

cat("H.M.of ll =",hm(ll),"H.M.of ph =",hm(ph),"H.M.of sy =",hm(sy))

The output will be:

H.M.of ll = 20.44167
H.M.of ph = 209.6673
H.M.of sy = 42.56131

Mode is the value of the series which is repeated for maximum no. of times i.e having highest frequency. A series can have more than one mode and can be called as bimodal (having two modes) series, trimodal (having three modes) series, multimodal (many modes) series.

In R, there is no inbuilt function for mode. We have to create a function for mode in the following manner.

#Create the mode function.

gmode= function(v) {

t = unique(v)

t[which.max(tabulate(match(v, t)))]

}

n= c(2,1,2,3,1,2,3,4,1,5,5,3,2,3)# ***Create the vector n with numbers.***

k =gmode(n)# ***Calculate the mode using the user function gmode.***

k # ***to show the value of mode of n variable***

The output will be:

[1] 2

cat("mode=",gmode(n)) # to print mode.= obtained value

mode= 2

gmode(ll) *# to show the mode of variable ll*

The output will be:

[1] 20.8

gmode(ph) *# to show the mode of variable ph*

The output will be:

[1] 213.1

gmode(sy) *# to show the mode of variable sy*

The output will be:

[1] 69.07

cat("mode=",gmode(ll)) *# to print mode.= obtained value*

The output will be:

mode= 20.8

cat("mode=",gmode(ph)) *# to print mode.= obtained value*

The output will be:

mode= 213.1

Median: Median is the middlemost value of the series. It is the value of the series of observation that occupies the middlemost position in the series after the observations of the series are arranged in order of magnitude.

The function in R for finding median is median() where the name of the variable containing the observations whose median is to be found out is written inside the bracket.

median(ll);median(ph);median(sy)

The output will be:

[1] 20.8

[1] 209.2

[1] 46.375

cat("Median of ll =",median(ll),"Median of ph =",median(ph),"Median of sy =",median(sy)) #

The output will be:

Median of ll = 20.8 Median of ph = 209.2 Median of sy = 46.375

Partition values: These are the values of a series of observations that divides the series into equal parts. The function in R used to find the quartiles is quantile(), where the name of the variable containing the series of observations is written inside the bracket.

To find out quartiles of the series

quantile(ll)

The output will be:

0%	*25%*	*50%*	*75%*	*100%*
16.300	*19.425*	*20.800*	*21.900*	*25.100*

quantile(ph)

The output will be:

0%	*25%*	*50%*	*75%*	*100%*
128.40	*194.40*	*209.20*	*237.05*	*330.03*

quantile(sy)

The output will be:

0%	*25%*	*50%*	*75%*	*100%*
21.1500	*34.3150*	*46.3750*	*58.4825*	*163.8000*

We can round off the result to our desired places of decimals by using round() function.

The syntax of the function is round(x,n)

The description parameters are :

x – outcome/value which is to be rounded off ; n – no. of places after decimal.

round(quantile(ll),2) # *will round off the result to two places after decimal*

The output will be:

0%	*25%*	*50%*	*75%*	*100%*
16.30	*19.43*	*20.80*	*21.90*	*25.10*

In the above output all the values are rounded off to two places of decimals.

If we want to find only the 1st, 2nd and 3rd quartile then we have to use seq() function inside the quantile function.

Illustration:

quantile(ll, seq(0.25,0.75,0.25))

*The first number is the minimum quartile, second number denotes the highest quartile and the third number denotes that we have to find the quartile of the series.

To get the quartiles for the variable ph

quantile(ph,seq(0.25,0.75,0.25))

The output will be:

25%	*50%*	*75%*
194.40	*209.20*	*237.05*

To get the quartiles for the variable sy

quantile(sy,seq(0.25,0.75,0.25))

The output will be:

25%	*50%*	*75%*
34.3150	*46.3750*	*58.4825*

To round off the quartile values of the variable sy up to 2 places of decimals

round(quantile(sy,seq(0.25,0.75,0.25)),2)

The output will be:

25%	*50%*	*75%*
34.31	*46.38*	*58.48*

To find the deciles we use the sequence operator function seq(0,1,0.1) inside the quantile() function.

get all deciles for the variables ll

quantile(ll,seq(0,1,0.1)) # get all deciles of the variable ll

The output will be:

0%	*10%*	*20%*	*30%*	*40%*	*50%*	*60%*	*70%*
16.300	*17.860*	*19.020*	*19.670*	*20.420*	*20.800*	*21.164*	*21.351*

80% 90% 100%

22.300 22.710 25.100

#To find the 4th decile for variable ll

quantile(ll,0.4)

The output will be:

40%

20.42

#To find the 4th decile to 6th decile for variable ll

quantile(ll,seq(0.4,0.6,0.1))

The output will be:

40% 50% 60%

20.420 20.800 21.164

To find all deciles of variable ph

quantile(ph,seq(0,1,0.1)) # get all deciles of the variable ph

The output will be:

0% 10% 20% 30% 40% 50% 60%

128.400 181.730 190.640 196.365 201.220 209.200 214.220

70% 80% 90% 100%

220.950 241.320 282.297 330.030

 # To get all deciles of variable ph rounded off upto 2 paces after decimal

round(quantile(ph,seq(0,1,0.1)),2)

The output will be:

0% 10% 20% 30% 40% 50% 60% 70%

128.40 181.73 190.64 196.36 201.22 209.20 214.22 220.95

80% 90% 100%

241.32 282.30 330.03

#To find the 6th decile for variable ph

quantile(ph,0.6) *# 0.6 can also be replaced with 3/5*

The output will be:

60%

214.22

#To find the 3rd decile to 7th decile for variable ph

quantile(ph,seq(0.3,7/10,0.1))

The output will be:

30% 40% 50% 60% 70%

196.365 201.220 209.200 214.220 220.950

To find the percentile

Illustration for finding 30th to 40th percentile for variable ph.

quantile(ph,seq(0.3,4/10,0.01))

The output will be:

30% 31% 32% 33% 34% 35%

196.3650 196.8075 197.4900 198.1725 198.7536 199.2840

36% 37% 38% 39% 40%

199.8136 200.3362 200.8588 201.1420 201.2200

rounding off the above result to one pace of decimal

round(quantile(ph,seq(0.3,4/10,0.01)),1)

The output will be:

30% 31% 32% 33% 34% 35% 36% 37% 38% 39%

196.4 196.8 197.5 198.2 198.8 199.3 199.8 200.3 200.9 201.1

40%

201.2

 # rounding off 3rd to 7th octile to one pace of decimal

round(quantile(ph,seq(3/8,7/8,0.125)),1)

The output will be:

37.5% 50% 62.5% 75% 87.5%

1.6 209.2 215.0 237.1 275.6

- Here 3/8 denotes 3rd octile, 7/8 denotes 7th octile, 0.125 denotes that we have to obtain octiles.

Range of a variable: Range of a variable is the difference between the maximum and minimum value of the variable. It is an absolute measure of variation. To find range of a variable first we have to find the maximum and minimum value of the variable. For this we have to use the function max() and min() where the name of the variable is to be written inside the parentheses.

Illustration to find range of variable ll:

max(ll)

The output will be:

[1] 25.1

min(ll)

The output will be:

[1] 16.3

range=max(ll)-min(ll)

range

The output will be:

[1] 8.8

cat("range of ll =",range)

The output will be:

range of ll = 8.8

#To find maximum value of variables ll, ph and sy

c(max(ll),max(ph),max(sy))

[1] 25.10 330.03 163.80

c(min(ll),min(ph),min(sy))

The output will be:

[1] 16.30 128.40 21.15

To find range of the three variables at a time.

c(max(ll)-min(ll),max(ph)-min(ph), max(sy)-min(sy))

The output will be:

[1] 8.80 201.63 142.65

c(cat("range of ll =",max(ll)-min(ll),"range of ph =",max(ph)-min(ph), "range of sy =",max(sy)-min(sy)))

The output will be:

range of ll = 8.8 range of ph = 201.63 range of sy = 142.65

Quartile Deviation: It is also a measure of dispersion used in case of qualitative variables which can be arranged in order of magnitude. It is half of the difference between third and first quartile.

$$\text{Q.D.} = \frac{1}{2}\ Q_3\text{-}Q_1)$$

QD1= (quantile(ll,0.75)-quantile(ll,0.25))/2

QD1

The output will be:

1.2375

cat("Quartile Deviation of ll =",QD1)

The output will be:

Quartile Deviation of ll = 1.2375

QD2=(quantile(ph,0.75)-quantile(ph,0.25))/2

QD2

The output will be:

21.325

cat("Quartile Deviation of ph =",QD2)

The output will be:

Quartile Deviation of ph = 21.325

QD3=(quantile(ph,0.75)-quantile(sy,0.25))/2

QD3

The output will be:

101.3675

cat("Quartile Deviation of sy=",QD3)

The output will be:

Quartile Deviation of sy= 101.3675

round(QD3,2)# *rounding off to two places after decimal*

The output will be:

101.37

Mean Absolute Deviation: It is arithmetic mean of the absolute deviation of the observations from their average (usually mean).

M.A.D $= \frac{1}{n}\sum_{i=1}^{n}|x_i - A|$; Where A is the mean, median or mode.

The function used to find Mean Absolute Deviation is MeanAD() where the name of the variable/vector containing the data are written inside the parentheses. The function MeanAD() is in the package DescTools. So to use this function we have to first install the package DescTools by clicking on install and then writing in the name of the package in the appropriate box of the opened window as given below:

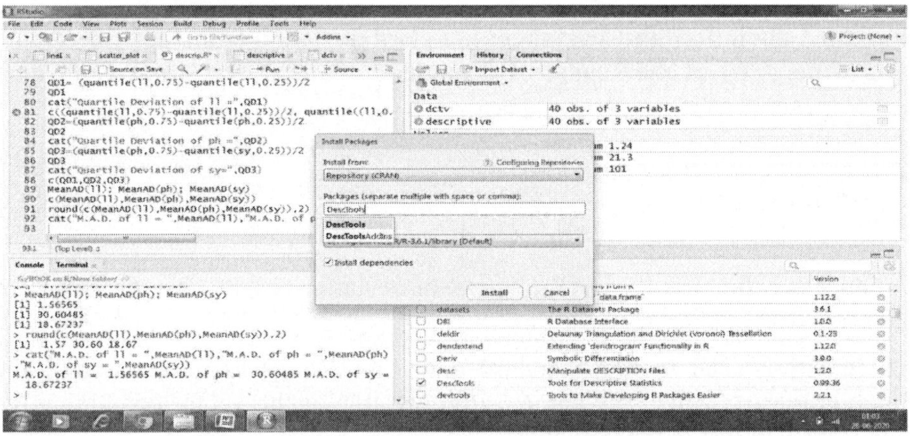

After the package is installed, it is to be loaded by the checking on the box near the name of the installed package.

#Illustration:

MeanAD(ll); MeanAD(ph); MeanAD(sy)

The output will be:

[1] 1.56565

[1] 30.60485

[1] 18.67237

#To find the mean deviation of the three variables at a time

c(MeanAD(ll),MeanAD(ph),MeanAD(sy))

The output will be:

[1] 1.56565 30.60485 18.67237

#Rounding off the result to two places after decimal

round(c(MeanAD(ll),MeanAD(ph),MeanAD(sy)),2)

The output will be:

[1] 1.57 30.60 18.67

cat("M.A.D. of ll = ",MeanAD(ll),"M.A.D. of ph = ",MeanAD(ph),"M.A.D. of sy = ",MeanAD(sy))

The output will be:

M.A.D. of ll = 1.56565 M.A.D. of ph = 30.60485 M.A.D. of sy = 18.67237

Standard deviation and variance: Standard deviation is the positive square root of the A.M of the square of deviations of the observations from their mean. Denoted as σ.

$\sigma = +\sqrt{\frac{1}{n}\sum_{i=1}^{n}\left(x_i - \overline{x}\right)^2}$; for individual data.

Square of S.D. is known as Variance. It is denoted by σ^2

Variance is defined as A.M of the square of deviations of the observations from their mean.

The function used to find standard deviation is sd() and for variance the function is var() where the name of the variable or the vector containing the data is written inside the parentheses.

sd(ll);sd(ph);sd(sy)

The output will be:

[1] 2.068924

[1] 41.70101

[1] 27.54888

var(ll);var(ph);var(sy)

The output will be:

[1] 4.280447

[1] 1738.974

[1] 758.9405

c(sd(ll),sd(ph),sd(sy))

The output will be:

[1] 2.068924 41.701008 27.548875

round(c(sd(ll),sd(ph),sd(sy)),2)

The output will be:

[1] 2.07 41.70 27.55

cat("Standard Deviation of ll= ",round(sd(ll),2))

The output will be:

Standard Deviation of ll= 2.07

cat("Variance of ll= ",round(var(ll),2))

The output will be:

Variance of ll= 4.28

cat("Standard Deviation of ph= ",round(sd(ph),2))

The output will be:

Standard Deviation of ph= 41.7

cat("Variance of ph= ",round(var(ph),2))

The output will be:

Variance of ph= 1738.97

cat("Standard Deviation of sy= ",round(sd(sy),2))

The output will be:

Standard Deviation of sy= 27.55

cat("Variance of sy= ",round(var(sy),2))

The output will be:

Variance of sy= 758.94

Coefficient of Variation (C.V.):

It is defined as standard deviation as a percentage of mean.

$$C.V = \frac{\sigma}{\bar{x}} \times 100$$

Uses of C.V.: Used to compare variability in two or more series.

 # Create a function for Coefficient of Variation

cv=function(x)

{return(sd(x)*100/mean(x))}

cv(ll);cv(ph);cv(sy)

The output will be:

[1] 10.02046

[1] 19.21615

[1] 53.17033

round(c(cv(ll),cv(ph),cv(sy)),2)

The output will be:

1] 10.02 19.22 53.17

#use of cat function to display C.V. = the value of cv for ll variable

cat("C.V. of ll =",round(cv(ll),2))

C.V. of ll = 10.02

One important thing to note is that since we have attached the data object named as 'dctv' in the beginning of this chapter by using , so we are able to use the variables ll, ph and sy in the data object dctv. But if the file dctv would not have been attached then to use the variables ll, ph and sy in the data object dctv we have use mention the name of the data object followed by $ before the name of variable. For e.g. to use the variable ll which is inside the data object dctv, we have to mention as dctv$ll

To detach the data object dctv, the function detach() is used where the name of the data object is written inside the parentheses.

detach(dctv)# will detach the data object 'dctv' which was already attached earlier.

cv(dctv$ll);cv(dctv$ph);cv(dctv$sy) *# to find CV of variable ll which is inside the data object dctv*

The output will be:

[1] 10.02046

[1] 19.21615

[1] 53.17033

cat("C.V. of ph =",round(cv(dctv$ph),2))

The output will be:

C.V. of ph = 19.22

cat("C.V. of sy =", round(cv(dctv$sy),2))

The output will be:

C.V. of sy = 53.17

Moments: The rth moment of a variable X about any point A is given by:

$\mu_r' = \frac{1}{n}\sum_{i=1}^{n}(x_i - A)^r$; for individual or simple series

where 'A' is any number. 'r' is the order of the moment.

If A= $\bar{x}$, then it is known as r^{th} central moment & is represented as μ_r .

If A = 0, it is called r^{th} raw moment about origin.

The function used to compute moment in R is moment(x,order,central)

Where,

x denotes the variable or vector

Order is the argument that takes the value 0 for zero order moment, 1 for first order moment, 2 for second order moment, 3 for third moment, 4 for fourth order moment and so on.

The argument central is a logical argument which is assigned TRUE for computing central moments and FALSE for raw moments.

Thus, moment(x,order=1,central=FALSE) gives the raw moment of first order for the variable/vector x.

The package for moment() function is moments. So to use moment function we have to install moments package and then load it.

moment(dctv$ll,order=0,central=TRUE)

The output will be:

[1] 1

moment(dctv$ll,order=1,central=TRUE)

The output will be:

[1] 1.776357e-15

moment(dctv$ll,order=2,central=TRUE)

The output will be:

[1] 4.173436

moment(dctv$ll,order=3,central=TRUE)

The output will be:

[1] 0.2925692

moment(dctv$ll,order=4,central=TRUE)

The output will be:

[1] 51.4713

moment(ll,order=0,central=FALSE)

The output will be:

[1] 1

moment(ll,order=1,central=FALSE)

The output will be:

[1] 20.647

moment(ll,order=2,central=FALSE)

The output will be:

[1] 430.472

moment(ll,order=3,central=FALSE)

The output will be:

[1] 9060.587

moment(ll,order=4,central=FALSE)

The output will be:

[1] 192480.9

moment(ph,order=0,central=FALSE)

The output will be:

[1] 1

moment(ph,order=1,central=FALSE)

The output will be:

[1] 217.0102

moment(ph,order=2,central=FALSE)

The output will be:

[1] 48788.95

moment(ph,order=3,central=FALSE)

The output will be:

[1] 11376683

moment(ph,order=4,central=FALSE)

The output will be:

[1] 2753374015

Skewness: Skewness means lack of symmetry. It determines the direction of variation in the observations of a variable.

The relative measure of skewness is given by $\beta_1 = \frac{\mu_3^2}{\mu_2^3}, \gamma_1 = \frac{\mu_3}{\mu_2^{3/2}}$; where μ_3 and μ_1 are third and first order central moments respectively.

$\beta1 = 0$ or $\gamma1 = 0$ indicates no skewness in the data.

$\beta1 \neq 0$ indicates there is skewness. If $\gamma1 > 0$ indicates positive skewness and $\gamma1 < 0$ indicates negative skewness.

The function skewness() is used for finding skewness in R where the name of variable or vector is written inside the parentheses.

Illustration:

skewness(ll);skewness(ph);skewness(sy)

The output will be:

[1] 0.03431531

[1] 0.760575

[1] 2.103666

Kurtosis: It means peakedness. It determines peakedness of the frequency curve.

β_2 &γ_2 provides measure of Kurtosis.$\beta_2=\mu_4/\mu_2^2$;$\gamma_2=\beta_2$-3

If the curve is more flatter, then it is called platykurticcurve.Here$\gamma_2<0;\beta_2<3$.

If the curve is more peaked, then it is called leptokurtic curve.Here$\gamma_2>0;\beta_2>3$.

If the curve is neither peaked nor flat it is called mesokurtic curve. Here $\gamma_2=0;\beta_2=3$

kurtosis(ll);kurtosis(ph);kurtosis(sy)

The output will be:

[1] 2.955137

[1] 3.620241

[1] 8.621776

The package for moment() function is moments. So to use moment function we have to install moments package and then load it.

Use of summary function: describe() is used to find the descriptive statistics of a variable where name of the variable/vector is written inside the parentheses. The package psych needs to be installed and loaded for using th describe function.

summary(ll)

The output will be:

Min. 1st Qu. Median Mean 3rd Qu.
Max.
16.30 19.43 20.80 20.65 21.90 25.10

Use of describe function to find the descriptive statistics

describe() is used to find the descriptive statistics of a variable where name of the variable/vector is written inside the parentheses. The package psych needs to be installed and loaded for using the describe function.

Illustration of describe function

describe(ll)

The output will be:

vars	*n*	*mean*	*sd*	*median*	*trimmed*
1	*40*	*20.65*	*2.07*	*20.8*	*20.64*

mad	*min*	*max*	*range*	*skew*	*kurtosis*	*se*
2	*16.3*	*25.1*	*18.8*	*0.03*	*-0.19*	*0.33*

Annexures

Annexure 1

Expenditure under different sports in a sports club

Sports	Expenditure (in Rs.)
cricket	720
hockey	360
tennis	120
football	240

Annexure 2

Profit (in Rs. lakh) made by a firm from 2016-17 to 2019-20

Year	ProfitA	ProfitB	ProfitC
2016-17	12	18	10
2017-18	16	24	20
2018-19	12	16	7
2019-20	20	12	18

Annexure 3

Production of two crops, crop A and cop B (in '000tonnes) of odisha for 30 years.

crop A	crop B
5942	4000
4759	4000
5868	4000
5329	4000
4047	4000
5894	4000
6857	4000
5856	4000
7140	4000
5842	4000
7104	4000

6829	4000
6730	4000
4779	4000
6602	4000
5769	4000
5612	4000
5029	6988
7536	7023
3587	6972
7114	7003
6964	7001
7427	7046
7432	7067
8346	7081
7639	7120
7705	7113
7770	7104
6695	7083
10362	7131

Annexure 4

Production of of a crop (in '000tonnes) of odisha for 30 years

Year	prod
1983-84	5942
1984-85	4759
1985-86	5868
1986-87	5329
1987-88	4047
1988-89	5894
1989-90	6857
1990-91	5856
1991-92	7140
1992-93	5842
1993-94	7104
1994-95	6829
1995-96	6730
1996-97	4779
1997-98	6602
1998-99	5769
1999-00	5612
2000-01	5029
2001-02	7536

2002-03	3587
2003-04	7114
2004-05	6964
2005-06	7427
2006-07	7432.12
2007-08	8346.1
2008-09	7639.81
2009-10	7705
2010-11	7770.49
2011-12	6695.24
2012-13	10362.03

Annexure 5

Production of two crops, crop A and cop B (in '000tonnes) of odisha for 30 years.

prod1	prod2
5942	5389
4759	4318
5868	5394
5329	4843
4047	3571
5894	5398
6857	6324
5856	5344
7140	6430
5842	5370
7104	6538
6829	6249
6730	6094
4779	4130
6602	6111
5769	5226
5612	4658
5029	4555
7536	6913
3587	3141
7114	6535
6964	6270
7427	6667
7432	6651
8346	7456
7640	6744

7705	6777
7771	6812
6695	5851
10362	9410

Annexure 6

Height (in cm) and Weight (in kg) of 30 persons

ht	wt
134	44
206	96
212	112
166	52
156	48
178	79
211	100
202	98
198	88
178	78
174	77
166	66
189	77
194	67
177	87
187	60
190	78
154	55
144	46
145	45
167	56
188	54
178	67
167	88
172	67
170	75
169	76
200	90
194	89
188	87

Annexure 7

Average Leaf length (ll), plant height(ph) and Straw yield (sy) of rice plots

ll	ph	sy
20.6	281.73	69.07
22.2	238.19	92.59
21.33	330.03	82.25
21.03	302.15	120.48
25.1	300.4	30.07
21.14	236.67	54.32
17.5	159.2	26.91
24.5	274.75	163.8
22.3	199.76	84.69
18.68	204.48	24.45
20.1	213.1	68.4
20.6	246.2	56.61
19.1	247.5	35.07
22.6	287.4	55.32
22.8	196.65	48.67
17.9	208.1	58.85
19.5	240.1	27.91
22.3	195.7	50.95
22.7	213.1	35.38
16.3	216.1	44.8
20.8	201.1	56.22
18.4	214.1	35.11
20.3	182.1	23.79
18.7	214.4	42.84
21.2	201.3	47.74
22.6	195.4	30.8
20.8	191.4	45.01
19.6	231.1	58.36
17.4	216.4	56.66
19.7	178.4	28.28
19.2	216.6	62.05
21.2	210.3	42.37
20.3	207.1	55.2
21.8	182.3	59.39
25.1	186.3	38.06
21.4	198.4	27.85
16.6	191.2	32.05
20.8	188.4	43.71
21.2	154.4	35.27
20.5	128.4	21.15